THE WISDOM OF SHEEP
& OTHER ANIMALS

Signed Edition

Rosamund Young

THE WISDOM OF SHEEP
& OTHER ANIMALS

by the same author

THE SECRET LIFE OF COWS

ROSAMUND YOUNG

THE
WISDOM
OF SHEEP
& OTHER
ANIMALS

*Observations from a
Family Farm*

faber

First published in 2023
by Faber & Faber Limited
The Bindery, 51 Hatton Garden
London ECIN 8HN

Typeset by Faber & Faber Limited
Printed and bound by CPI Group (UK) Ltd, Croydon CR0 4YY

Illustrations © Joanna Lisowiec, 2023

*Every effort has been made to trace copyright holders and
to obtain permission for the use of copyright material. The
publisher would be pleased to rectify any omissions that are
brought to its attention at the earliest opportunity*

A CIP record for this book
is available from the British Library

ISBN 978-0-571-36825-9

Printed and bound in the UK on FSC® certified paper in line with our continuing
commitment to ethical business practices, sustainability and the environment.
For further information see faber.co.uk/environmental-policy

2 4 6 8 10 9 7 5 3 1

Why does the lamb love Mary so?
    The eager children cry;
Why, Mary loves the lamb, you know,
    The teacher did reply.

– Sara Josepha Hale, from
Poems for Our Children

CONTENTS

INTRODUCTION
The Journey to Kite's Nest

It is the winter of 1947 and the village where I will be born in six years' time is completely cut off from the outside world. The drifts are fifteen feet high, and my father and grandfather have worked all morning to dig a tunnel through the snow across the road from the farmhouse to the farmyard in order to milk the house cows and feed the sheep. My grandmother collects snow in every available receptacle to melt for water as all pipes are frozen.

According to Grandpa's journal, the first snow starts to fall on 18 January. It stays till the middle of April. It is many weeks before some of the village men manage to walk to the nearest town, but they return with just nine loaves. In the interim, my grandmother makes awful pastry with coarsely milled flour and lard; she doesn't attempt bread. My mother, just twenty years old, develops pneumonia and Doctor King skis several miles across the fields to treat her.

Once a day, whenever possible, the heavily pregnant ewes are encouraged to walk round the perimeter of the unfenced field known as Witchcraft, to give them the exercise they need to stay healthy. When lambing gets under way my grandfather brings two orphan lambs, Sally and David, to the house, to be reared by my mother and her younger sister. Only in such exceptional weather would my grandfather ever have permitted 'pets' of any kind in the house. A fire is lit in the sitting room every afternoon and allowed to die down slowly before bedtime. Sally investigates the embers and, as soon as she deems them sufficiently safe, lies down on top of them and stays there till daybreak. David curls up where the brown woollen blanket that hangs inside the blue-velvet door curtain gathers in a pool on the floor. Both of them become totally house-trained. They are fed on fresh cow's milk and they thrive.

As soon as they can venture outside, the two lambs invent a game they never tire of playing: bounding up the snow-covered road that separates the farmhouse and farmyard, climbing the stone granary steps, leaping off the top and jump, jump, jumping back to the garden gate to start again.

My mother's father was thirty-nine when she was born and she was introduced to every aspect of farming from a very young age. When she was eighteen months old her sister arrived, followed six years later by her brother. A son to take over the farm was considered essential, and he was deemed the most important member of the family by everyone including his sisters.

My mother became a lonely child, sent for extended 'holidays' to her paternal grandparents from the age of two. It was a daunting time; her grandmother was very kind, but seriously overworked and under-appreciated, and her grandfather authoritarian and difficult. He insisted that my mother eat everything on her plate, even if it took her two or three mealtimes. She later believed that it was the generally poor and unappetising food she received, both there and at home, that set her on the path to lifelong ill-health. From the age of four, she found an escape of sorts in reading. It seems likely that her family, like the theatre director Joan Littlewood's, would have thought that 'to be found reading would be worse than lying in bed all day'.

In those days, everyone who lived on a farm became involved in the daily and seasonal round of work. When my mother was ten, her father told her to take a load of hay home. On asking how she would navigate the frighteningly narrow river bridge, she was assured the horse would 'know', and of course it did. My grandfather had been deeply affected at having been ordered to abandon his horse during the First World War, when the Warwickshire Yeomanry was disbanded and amalgamated with a foot regiment. According to my mother, as compensation he received just ten shillings (i.e. 50p, worth about £30 in today's money). His devotion to horses never waned: he never learned to drive a car or tractor. Grandpa died unexpectedly when I was a baby, but my brother Richard was three and he remembers sitting on Grandpa's knee and being given the reins to hold when driving through the village in a horse and cart.

———

My father was a first-generation farmer, inspired by a schoolfriend on whose father's farm he learned to milk cows and love farming. His first job, at the age of fifteen, was delivering milk in seventeen-gallon churns by pony and trap in the town where he was born, doling out measures to the women who came out carrying jugs. They sometimes brought him freshly made cakes, and he loved every minute. After a tough apprenticeship on a Northumberland farm, he left his native county of Durham while still in his teens to milk the small herd of Dairy Shorthorns on the farm of my grandmother's cousin, Jack Hodges, in Warwickshire. My mother, who was the same age and lived nearby, was invited to supper to meet him. As she leaned her bicycle against the wall, she saw my father's kindness towards the cows and decided there and then that she would marry him. She wanted to spend her life with someone who cared about all animals. He insisted on cycling home with her later, lending her his scarf as the evening was chilly.

I was born in June 1953 in the cottage on my grandfather's farm in the village of Condicote, but we moved when I was less than two weeks old to a remote smallholding in Clapton-on-the-Hill, rented from Gloucestershire County Council. It was truly off the beaten track but in the early evening of the first day, thirteen cats appeared on the doorstep asking to be fed! The fields were steep and the soil was heavy and waterlogged. It had been occupied since 1919 by an invalided First World War veteran. My father set about draining the wettest fields, digging almost a mile of trenches by hand with a ditching spade, laying red clay pipes in the bottom of the trenches and then back-filling with soil. We lived in a bungalow in the middle of nowhere, several hundred yards from the farm buildings, with magnificent views but no telephone, electricity, or easily accessible water supply, just a temperamental pump in the back kitchen. Within a few months, though, a telephone line and mains water were connected. Electricity was laid on too but the council wouldn't pay to wire the bungalow, so my father did it himself.

There is a grainy photograph of me aged around two years, bottle-feeding a lamb in my grandmother's orchard. It would be more than half a century before I was able to decide to keep sheep myself. The intervening years were spent with cows and pigs and hens. 'Other people' kept sheep and the only ones I ever saw were in huge flocks. I'm ashamed to say that I believed what I was so often told, that they didn't have individual personalities.

Cows were not just part of our life but the very core of it. I helped to move the electric fence most days when I was not at school and tested its strength by holding a blade of grass to the wire, which gave me a small shock rather than the powerful one I occasionally got by accidentally touching the wire itself. I roamed freely among the cows from the time I could walk. Only Susan was a bit inclined to toss her head, so we always walked behind rather than in front of her. I frequently picnicked in fields of cows, or sunbathed, eyes closed, without the tiniest worry that they would harm me. They often surrounded me, snuffling at me with grassy breath and licking my wellington boots.

My days were filled with the routine of school, before the excited rush home to be on the farm. One of my jobs was bucket-rearing calves. We milked a herd of pedigree Ayrshires. The calves were kept separately and fed whole milk twice a day. I would carry four small plastic buckets each containing four pints of precisely warmed milk, two on each arm to leave my hands free for opening latches. The trick was to time my entry to the calves' home and swiftly position the buckets under each nose without any being spilled. I had of course watched this being done countless times before attempting it myself, and I knew that to spill any would not merely be expensive but would cause havoc. Four calves each receiving their allocation simultaneously spelled success and contentment. If one bucket or part thereof was spilled, the deprived calf would try to steal from one of the others, resulting in much if not all the milk being lost, and the whole procedure having to be repeated.

Our Scottish Border Collie, Roy, was exactly the same age as me. He was bought by my father, by telephone, from a breeder he knew. At the age of six months Roy travelled unaccompanied in the guard's van, all the way from Scotland to the old railway station in Bourton-on-the-Water, on the long-gone branch line from Kingham to Cheltenham. This journey involved many changes of train, and we were all waiting anxiously on the platform. As soon as Roy alighted, my father called him by name and the dog ran straight to him. Evidently the various guards had all made a fuss of him because he seemed extremely happy.

Roy was fully trained – he responded to Scottish commands – and happily worked with any of us. He could distinguish between calves and cows in adjacent fields if we said the words clearly. He had a sweet temperament and we all loved him. But he was used only twice a day to bring the cows in for milking, up very steep fields on our first, forty-seven-acre smallholding. The rest of the day was his own. He enjoyed our company and could always be seen in family photographs, even when we had no knowledge he was there until the picture was developed. Every time he was given a bone he would immediately hurry away and bury it, but he could never remember where he put it. We had to get into the habit of secretly following him so that when he started looking for it we could tactfully show him where to dig.

From the time I was about six, Roy and I would walk along the drive between the milking parlour and our bungalow, lined with alternate bushes of double-white and purple lilac interspersed with laburnum, whose poisonous qualities are belied by its outward beauty. On and on we would walk, to whichever field the cows were grazing. On what today seems like a very small farm, the distance felt long to my short legs. Then I would stand at the top of one of the all-steep fields and call: *cuuuu'p, cuuuu'p, cuuuu'p, cuuum'-on,* which would make some of the cows look up. I would say the magic Scottish command to Roy and he'd be away, but he always slowed down once he'd rounded the cows up, and walked them home gently.

———

My tenth year, 1963, was the coldest for two hundred years. I enjoyed it immensely. The snowdrifts were beautiful and excitingly challenging. I had a hand-me-down teddy-bear suit, sent by an American relative, which zipped me into a warm and draught-free onesie. The walk from our house to the farmyard required the skill and courage of an Arctic explorer, which is what in my mind I became.

The trials my parents had to overcome that winter were huge. Frequently there would be no electricity for long periods of time, but my resourceful father managed to attach belts from the milking machine to the tractor pulley he normally used to saw firewood. With the pulsators connected to the tractor battery he managed to milk our herd of Ayrshires twice a day as usual.

Our neighbour, Frank Moy, took a three-legged stool and set to work to milk his Dairy Shorthorns by hand. He just had time for a very strong cup of coffee after finishing the morning milking before commencing on the afternoon stint. I learned more from his empathy with his cows and deep knowledge of most things than from any other person. Frank demonstrated to me just how much cows can be trusted when they are never hurried or bullied. He walked his Shorthorns in for milking through part of the village, winding slowly downhill to his farmyard. He opened the gate from their field and then followed behind them. They knew where to go and behaved beautifully, walking into the traditional milking shed and stationing themselves in their own individual stalls, ready to be fed and milked.

Getting the cows milked and the milk into ten-gallon churns in that cold winter was one thing, but taking it to rendezvous with the lorry that could no longer drive to the village because of the ice and snow was an altogether different and more dangerous adventure. In fact, for the first six weeks of Arctic weather, the milk had to be tipped away. By mid-February, my father, Frank and other men from the village had forged a passageway by shovel and fore-loader through one long section of snowdrifts, so that our tractor and transport box could slide down a very steep lane where the snow had not drifted, to meet the milk lorry on the main road each day. Richard or I sat in the transport box to make sure the churns of milk did not tip over. On the way back the tractor would gather speed in an attempt to get up the steep icy slope. Often we didn't quite make it to the top, and had to shovel more grit onto the road before having another go. After this, it became possible to bring in supplies of fresh food: the first for six weeks. I wonder how many people could manage without fresh supplies for six weeks today? It wasn't easy then and everyone went short in one way or another but it was a different world. None of us had heard of a supermarket, let alone seen one. Most people had some home-grown food stored away and went shopping only occasionally.

When summer finally arrived, our tiny, remote village organised a fete, the highlight of which was to be a hundred-yard dash, open to all ages, with an alluring cash prize of £5 that proved irresistible to a group of fit young men from surrounding villages. Richard, aged fourteen and wearing his shorts, running vest and prized spiked running shoes, was a contender. My thirty-seven-year-old father arrived a few seconds before the start, still wearing wellingtons and his cap, and outran them all.

———

Richard had steered a tractor at the age of four and could plough quite well, with a two-furrow plough, by the time he was seven. But before I could enjoy baling or mowing or any other tractor-oriented activity, I had to endure an interminable wait for my thirteenth birthday, because a law change in 1958 prevented me from driving or even riding on a tractor before then.

One of my first jobs on my parents' farm was unforgettable: driving an ancient Fordson Major tractor, long before the days of cabs or roll bars, the soft wind blowing the scent of glorious wildflowers – moon daisies, sainfoin, vetches, sorrel, quaking grass and scabious – as I mowed with a reciprocating knife, leaving neat rows of sublime colour behind me.

My early life was punctuated by wildflowers. Lady's smock and several different clovers in the fields near our home, scabious lining the roadside on the way to primary school, violets and primroses edging the bare earth each side of the footpaths. As winter eased into spring I looked forward to each one starring in turn.

Flail mowers had not been invented and nature graded her own borders with the tallest plants nearest the hedge and the shortest by the side of the road or track. Not only were the roadside verges a glorious carpet of colour, but the absence of mechanical hedge-cutters meant the accompanying hedges were delicious, nutritious, thick, often impenetrably safe larders and nesting sites for wild birds. They were tall too, providing high-rise options as well as low-level accommodation. Gaps in hedges often had to be hurriedly filled with whatever came to hand. One nursery rhyme tells us:

The man in the moon was caught in a trap,
For stealing the thorns from another man's gap.

During my school years, farmers were given financial inducements to create larger fields by destroying hedges and ditches, and (as soon as such methods became available) to minimise the hedges that remained by frequent, mechanical cutting. The value of a field-grown crop was the only calculation used, and it was considered commendable to steal every possible extra inch from the margins by removing overhanging branches or flower-rich edges. The intrinsic value of a hedge to provide shade, stabilise soil and control water run-off as well as to provide habitat for natural predators of crop pests was ignored – now we know that they sequester and store carbon too. As Nicolas Lampkin noted in his seminal 1990 book *Organic Farming*: 'The increased yield of crops within a hedged field more than compensates for the loss of yield in the immediate vicinity of a hedge.'

Once I had learned to drive a tractor, the next step was to take a test. I was sixteen, and Richard said I would be more use if I could drive on the road. He was now farming a rented farm eleven miles away, and often moving machinery and straw between the two farms. I had only about twenty yards to drive to meet the examiner but I had to tie L-plates on. Everyone in the village decided to watch, while pretending to be busy doing something else. The examiner said he was going to hide behind a bush on the side of the road and when he jumped out, I was to perform an emergency stop, but as the bush was too small to conceal him, I had plenty of warning. He then asked me to drive 'at my normal driving speed' along the main A46 (now B4632) and do a three-point turn in the first gateway. I was too nervous to admit I was still in low range – normally engaged for slow reversing and manoeuvring rather than road driving – and by the time I reached the gate, a car had overtaken me and parked there. I carried on and turned in the next gap, which happened to be out of sight of the examiner. By the time I next emerged into view, he had almost given up hope of ever catching sight of me again. I passed, largely because I don't think he could face the idea of a repeat performance.

We moved from Middle Hill Farm, Saintbury, to Kite's Nest near Broadway in August 1980. Moving house is one thing, but it's got nothing on moving farm, something that isn't to be undertaken lightly. We fell in love with the farm as soon as we saw it, tucked into a valley on the scarp slope of the Cotswolds, with the land rising from about three hundred feet above sea level to almost nine hundred feet at the highest point. As a result it's not the easiest farm to manage, but apart from its fascinating contours, showing signs of past geological activity and human imprints, what appealed to us was its stunning natural beauty and variety, and the diversity of plants and wild animals that came with it.

The soils range from heavy clays through alluvial loams to light, stony Cotswold brash, with small areas of both peat and sand. Previous owners have planted a wide variety of different trees, each suited to the specific soil types. There are willow, poplar and alder in the lowest-lying areas; walnuts, oak, ash, wild cherry and silver birch on the better-draining deep soils; field maple, sycamore and more ash at four to six hundred feet; and the shallow-rooting larch and beech trees ideally suited to the thin, more typically Cotswold soils higher up.

The main enterprise in the early decades at Kite's Nest was the beef suckler herd, in which the cattle stay in family groups for life. I hankered after sheep, but the daily vigilance they require, including round-the-clock attention during lambing, was impossible while I was caring for Mum. Always in extremely fragile health, she became utterly dependent on Richard and me in the last years of her life. Much of our farming was done by torchlight in the stolen hours after midnight, and the cattle got quite used to our nocturnal routines. Sheep, however, had to wait.

We now have two flocks: Lleyn and Shetland. I'm writing this on the last day of lambing, and on the farm there are over three hundred sheep.

———

To begin with, I stored away everything that happened on the farm in my memory, putting down very little in tangible diary format. My earliest memories were the most vivid, and it did not occur to me that I might ever forget anything. In the late 1990s, the farm attracted a lot of media interest as the public's interest in organic food surged, and one visiting journalist asked particularly probing questions which inspired me to embellish my answers with some of these recollections. Each time a 'story' ended, he asked if I had any more. When I admitted I did and that new ones happened every day, he suggested that I consider writing them down. That same day, I started a notebook-cum-diary and found to my surprise that I thoroughly enjoyed recalling and recording daily farm events.

This book is based on that diary, kept in the kitchen-table drawer and scribbled in after meals. But there were so many days and even years with blank pages when actual live dramas swallowed up every minute of each day that much here is transcribed from my book of memory after all. As you read these pages, new stories are unfolding. The months and the seasons follow their unceasing rhythm, but every day the animals and the farm itself stop me in my tracks with surprise.

A COLD LATE WINTER

Kite's Nest had one significant disadvantage when we first came here: a very limited range of buildings. Most of them were in poor condition and none of them were really suited to modern farming, being either too low or too narrow for tractors.

On the farm that Richard rented when he left school, he had had no trouble obtaining capital grants of 40 per cent for a range of buildings to keep all the cattle comfortable in adverse weather and all the hay, straw and farm machinery under cover. In the early 1970s, he was still using nitrogen fertiliser and other chemicals. I'm not sure if it was specifically stated in the legislation, but these were effectively the key to obtaining the grants, because the money was conditional on increasing yields, and also on specialising in preferably one or at most two enterprises.

When we applied for a grant at Kite's Nest, we got a shock. We were told that we were not eligible as we wouldn't be able to meet the productivity criteria without nitrogen fertiliser. The Agricultural Development and Advisory Service (ADAS) was, at that time, a free service for farmers. One very kind ADAS officer took up our case, which was that by using clover in our grasslands we could increase the farm's output without nitrogen. But the

Ministry of Agriculture, Fisheries and Food (MAFF) was institutionally hostile to the concept of organic farming and totally committed to intensification through increasing use of agrochemicals. It took the ADAS officer five years of battling on our behalf to get MAFF to agree reluctantly that we could apply. He said he almost lost his job as a result.

It seemed as if everything was against us at the time: interest rates were soaring and the cost of new buildings had increased so much that even with the grant they were going to cost a lot more than they would have done had we put them up when we arrived in 1980. The grant scheme was also being phased out, limiting our time to arrange finance and get planning approval, which we needed because the farmhouse is listed.

The scale of the problem posed by the condition of the farm's buildings became clear in the first months of 1983, which saw a very cold late winter with biting east winds against which we had to build temporary windbreaks with straw bales to keep the cattle warm enough. Come April it seemed as if spring had finally arrived, but the weather suddenly turned cold again and, very unexpectedly, snow was forecast. We got the cattle back in and bedded them down. Rather than the sprinkling of snow we'd expected, we woke up to find six to eight inches of wet and frozen snow lying everywhere, including on the flat roof of the flimsy pole barn erected by our predecessor.

Very lightweight plastic sheets were supported on beams only two inches wide. This was enough to support the sheets, but completely inadequate to take the weight of the snow as well. There were close to a hundred cattle in this barn when the roof caved in during the night. Clearly there must have been some creaking before the collapse, as the cattle had all huddled at one end, where about a quarter of the roof was still precariously in place. Miraculously none of them were hurt.

Another heavy flurry of snow left me wondering how to rescue our tiny flock of sheep, who I knew would be huddling against the top wall of the Seven-Acre Field. Richard hitched our old cattle transporter to the tractor, and we set off uphill on the impassable-looking farm track. I clung on precariously as he drove in blizzard conditions, zigzagging and sliding. Up one very steep part, the wheels spun so much we almost came to a stop, but we just managed to reach the top track and with it the flatter ground. When we reached the sheep, we opened the transporter and fought against the wind to push the gate open just far enough for them to squeeze through. Although they had never travelled in any vehicle before, they all hurried inside and we tobogganed terrifyingly home.

Fortunately, the snow didn't last long and we were able to turn the cattle out and feed them in the field. The pole barn was a complete write-off. We'd tried to get it insured, but every company had balked at its flimsiness. Not only were there broken beams and large pieces of plastic roofing sheet everywhere, but many of the sheets had shattered into hundreds of small pieces. These all had to be picked out of the straw by hand before we could clean out the three-foot-deep winter bedding. Without a grant we couldn't afford to replace the barn with a new building, but we had to have shelter ready for the following winter. Richard decided the only option was to upgrade the size of the beams significantly and re-roof it himself with second-hand corrugated sheets. Little did we know at the time, but this would have to suffice for the next thirty-three years.

THE FOUR SEASONS

Spring is so brief. It is glorious relief, after a long winter of laborious and relentless feeding, to be able to let the cattle graze grass in abundance instead of hay or silage. No sooner do they have the right to roam freely, however, than we need to confine them to certain fields unsuitable for haymaking and leave the rest of the grass to grow.

Summer is all about winter: assessing the weather, mowing, spreading, rowing up, baling and carting the precious fodder to the safety of the barn. Farmers have to plan a long time ahead to be sure they have sufficient feed for their animals, just as people always had to grow their own food and learn how to preserve as much as they could before the advent of supermarkets.

Autumn will vary. If you grow crops you will be totally absorbed with the harvest and preparing the soil for planting. If you have only livestock, there will be time to enjoy them enjoying just being. The tough work of winter will seem a long way off, and as your animals eat their way towards it, grazing the grass as it declines in abundance and quality and grows more slowly, you will hope and almost believe that autumn will stay kind until spring.

Winter is eating the summer bounty, spiking and unzipping the bales, repairing fences when you can find time on a dry day, and cleaning, sharpening, oiling, greasing and servicing the kit ready for summer.

In many ways, sheep are less trouble than cattle in winter. They need to be fed of course but they are lighter and don't poach the pasture into deep mud with their hooves as cattle do when it is wet. All of our animals prefer to be outside if at all possible, but the cattle do appreciate the shelter of the barns in adverse weather while our sheep absolutely hate being kept inside and make their feelings very clear. Their coats are weatherproof, coated with lanolin to repel water and able to protect them from fierce winds.

LONDON, 1899

I knew eleven of my thirteen great-aunts and uncles – my grandparents' brothers and sisters – and remember stories of milking and feeding cattle, of water wheels and immensely heavy sacks of flour, and of special meals being cooked daily to feed the prized greyhounds and hens. One great-uncle's only trip to London in his eighty-five years was at the age of eleven, in 1899, when he was sent there by train with two greyhounds to deliver to a customer. Apart from the fare, he was given sixpence, but rather than buy food he spent it all on a souvenir ashtray of the Tower of London so that he could prove he had visited the capital and savour the memory. We have it still.

A ROYAL VISITOR

On 1 August 1989, the Prince of Wales visited Kite's Nest on his first official visit to an organic farm. We enjoyed the day immensely. We drove round the farm in our old, battered (but polished) Range Rover and, slowing down through one gateway, HRH wound down the window to stroke a heifer called Catherine. I was just about to inform him that she wasn't friendly and never allowed anyone to stroke her, but instead I watched in silence as she stood perfectly still while he did just that.

The visit had been planned over the previous nine months, as an event at which the Prince would announce that, after experimenting on eighty acres, he had decided to convert the whole of the Highgrove estate to organic methods. While our farm was chosen for the visit, and the official line was that we'd been one of the farms to influence him, it was really a tribute to the whole organic movement.

The Lord Lieutenant of Worcestershire came to see us a week beforehand and asked the most searching, interesting and amusing questions and, of course, the police spent a great deal of time and energy making sure the security was well thought out. There was an unexpected short shower first thing; the rest of the day was perfect weather. The police sniffer dogs jumped in the Range Rover with muddy paws so we quickly had to remove the seat covers and expose the worn seats we had cleverly hidden.

The gleaming red helicopter landed in one of our few flat(ish) fields. After driving round the whole farm with Richard and me, admiring many of our rarest wildflowers and talking about every topic under the sun, HRH joined us for lunch – all home-grown and organic – on the lawn with my parents. The Prince was not told that my father had been unwell, nor that he was very hard of hearing, but he seemed to sense this and spoke directly to him, drawing him into conversations he might otherwise have missed. Mum, of course, discussed literature, politics, religion and everything else – or, rather, she talked about them and the Prince listened!

BODY LANGUAGE

Sheep know how to ask for help, using sign language in the form of body position and eye contact, with an occasional baa that can be surprisingly deep-voiced.

Nessie was ready to give birth to triplets but the first two were trying to be born backwards and at the same time, so she failed to make any progress. She watched the Land Rover drive into the field, walked straight over to my door and looked at me. She didn't look exhausted (though she probably was), she wasn't pleading, she didn't look distressed. She just said quite plainly that she'd tried to lamb and couldn't and she needed help, right now. It was a calm sheep stare speaking volumes, and every shepherd in the world would have understood exactly what it meant. It was quite difficult to work out which head and legs belonged to each other and it certainly felt as if she would never have managed on her own. She put her trust in us totally and we were so glad to be of use.

Even the least friendly sheep will ask for help, though as a species they are very brave and skilled at not appearing to be in pain and thus vulnerable to predators. Some individuals will go to enormous lengths to avoid people, so when we can see that they won't manage alone, we have to resort to whatever tactics we feel are necessary in order to outwit and frequently outrun them.

Lambs don't take long to grow up; the first intake of milk has a miraculous effect. They will conveniently wag their tails as soon as they latch on to a teat, and shortly after drinking they usually attempt a skip or jump. After a short nap and another drink they are ready for whatever the weather throws at them. If there is a huge downpour at the moment of birth, however, that can put everything out of kilter. Most lambs will cope, especially if the ewe has had lambs before and knows what she's doing. If we are on hand and have a spare Land Rover canopy (we own seven) we will bring it to create an instant, temporary shelter, for which the ewe is invariably grateful.

LOVE OF BIRDS

Each year, the arrival of autumn sees birds crowding the bird table: great spotted woodpeckers, blue tits, great tits, dunnocks, marsh tits, willow tits, coal tits, robins and goldfinches. One day a squirrel appears, climbs the wooden pole and starts eating the wheat, the apple, the pear. I open the window to shoo it away. It takes no notice. I shout at it. It ignores me.

I know the squirrels have a huge store of walnuts; day after day, I watch them picking and transporting the nuts to a hollow poplar tree. Even though grey squirrels are a pest species introduced from North America in the 1870s, and have been responsible for the dramatic decline of the red squirrel I so loved to see when I was a child, I don't begrudge them their winter store. But the bird table should be sacrosanct.

I am watching from the kitchen window, holding a basket for some reason, and I keep it over my arm as I walk outside and tell the squirrel to go away. I place the basket on the ground and before I can take a step, the squirrel walks towards me, looks up appealingly and jumps in. I am suitably amazed but feel there is no point in saying so. I pick up the basket and take the squirrel for a short walk round part of the garden. I return to the same spot and place the basket on the ground. The squirrel jumps out.

The next day I put extra food on the bird table and some on the window ledge, and leave the window open while I eat my breakfast. The squirrel appears, sits on the windowsill, sizes up an apple and looks at me. He puts his hands round the apple but looks disappointed. I fetch a smaller apple and place it close to him. He picks it up and starts to eat, turning it round and round as he nibbles. He looks happy. He eats a small piece of bread, a small pear, then a few mouthfuls of wheat.

He comes every day and I stroke him gently on the back of his head and under his chin. When he wants me to stop he grabs my finger. I stroke him with only one finger. I stop. He lets go. Every day he eats a selection of the food I offer, sitting on the windowsill inside the kitchen. When he has eaten enough he plays on the lawn, rolling over like a kitten.

After several weeks, a day arrives when he doesn't visit. The next day he comes with a torn ear and a scratched face. He doesn't come again for nearly a week. The other ear has been torn. He plays on the lawn as before and eats everything I have put out for him. He does not come again.

The following spring, while I was driving Mum along the top edge of the Oxstalls Field, we saw a young tawny owl on a branch of a very old oak tree. We stopped and watched him as he tried to stay awake but kept dozing off, eyes closed, then half open, then closed again. After a while, we reversed away from the tree and came home to get Dad. The four main forces that held my parents together were animal welfare, love of birds, classical music and the commonality of their money. Mum stayed at home while I drove Dad to see the owl. He was beyond delighted. He had already suffered two strokes but was feeling well and happy and, as always, finding ways to be incredibly useful. We stayed, ogling the baby owl for ages.

As it would turn out, Dad had only three months to live, though luckily none of us had any inkling of that then. He died on 8 June 1993 at the age of sixty-six. It was late, so I took a cup of tea upstairs and found him lying in bed looking peaceful. He had gone up slightly earlier than usual as Mum had told him that Chopin would be on Radio 3. The radio was still playing.

Les Dawson and James Hunt died later that week. My father would have appreciated being in the company of two people he greatly admired.

For my fortieth birthday, a few days later, I was given a high-quality tape recorder and microphone that he, Mum and Richard had jointly chosen, and I proceeded to record birdsong in many locations at dawn and dusk. I drove Mum round the farm in the old Range Rover and whenever we travelled to a remote corner to place the recorder, we inevitably recorded ourselves driving away from it. One day when a new customer, who we later realised was a keen ornithologist, was invited to come with us to see the cows, he asked if we had any hobbies. Mum misunderstood the question and explained in a beautifully jumbled way that yes, we liked to take our recorder to remote locations on the farm and, well . . . record ourselves driving home.

Wagtail and Baby

A baby watched a ford, whereto
  A wagtail came for drinking;
A blaring bull went wading through,
  The wagtail showed no shrinking.

A stallion splashed his way across,
  The birdie nearly sinking;
He gave his plumes a twitch and toss,
  And held his own unblinking.

Next saw the baby round the spot
  A mongrel slowly slinking;
The wagtail gazed, but faltered not
  In dip and sip and prinking.

A perfect gentleman then neared;
  The wagtail, in a winking,
With terror rose and disappeared;
  The baby fell a-thinking.

– Thomas Hardy

BSE STRIKES

The 1990s were dominated by the spectre of Bovine Spongiform Encephalopathy (BSE). The decade began with Agriculture Minister John Selwyn Gummer attracting major media coverage by trying to get his young daughter to eat a beefburger in front of the cameras to assure the public that beef was safe, despite the widespread belief that BSE in cattle was the cause of the growing number of young people dying from a new form of Creutzfeldt-Jakob disease. The stunt backfired. She refused to eat the burger, as it was too hot, and the publicity increased public distrust.

In March 1996 the Secretary of State for Health, Stephen Dorrell, informed parliament that BSE was, as suspected, the cause of new-variant Creutzfeldt-Jakob disease, and the EU imposed a worldwide ban on exports of British beef lasting until May 2006. Initially we felt this would not affect us; we had a closed herd and had never fed meat and bone meal, which was widely thought to have caused the disease. Demand for beef plummeted nationally but our sales remained stable. I made a brief appearance on *Newsnight* to be interviewed about our situation.

The following month, all cattle over thirty months of age were banned from the food chain. This affected us massively, effectively halving our profit overnight, as we had been making pure steak burgers from the older cows since 1984 and had built up a good trade, selling to fourteen small wholefood shops as well as the retail customers who came to the farm. This trade completely disappeared. The Soil Association asked for an exception to be made for organic farmers but this was turned down. Richard wrote to Douglas Hogg, then Minister for Agriculture, asking if we could be granted an exemption and explaining why. The letter was never answered.

In 1998 our MP, Peter Luff, presented a petition to parliament on behalf of 523 of our customers who had signed to say that they wished to be allowed to buy meat from our cattle that were older than thirty months. We felt sure of the merits of our case and were genuinely optimistic but it had no impact on the government. Tony, our butcher, now had far less to do and we had to ask him to spend half of his time as a general farm worker.

WHEN THE WIND BLOWS

My mother and I have both just read *When the Wind Blows* by Raymond Briggs.

Me: The world will end in four minutes exactly.

Mum: Well then, you've just got time to sweep the kitchen floor.

Me: Must I? There'll be no kitchen floor in three minutes.

Mum: I know, but someone might survive and whatever would they think?

WORDS OF NO SYLLABLES

Our lambs seem to be developing new ways of communicating with us. One morning Richard drove Mum to see the sheep. One of them, four-month-old Tilly, greeted them, then walked over to yesterday's now chewed and leafless willow sticks, picked one up and shook it. Mum asked Rich to fetch some more.

And it's not just the lambs. One summer evening I went to see Giselle, to find her lying comfortably with her sore teat uppermost, so I was able to apply udder cream without disturbing her at all. I did not speak, of course; I've learned better. The old Red Bull was lying next to her, so I seized my chance to check all four of his feet while his weight was nicely distributed along his recumbent length. As soon as I began, Giselle started to complain, moaning quite loudly. I returned to her and immediately saw a short, hard stick, tightly wedged between the clays of one front hoof. I carefully prised it out and she was silent.

Earlier that evening, Tony, for whose sole benefit we receive letters addressed to 'The Wages Supervisor', had come, smiling, to the door with an account of his own. He had walked the house cows in as usual but had not noticed an extra one, a look-alike, until they were in the barn. He began an enquiry and soon noticed a piece of string wound round her back leg, which he promptly removed.

'That's obviously why she came down,' he said. 'They're not daft.'

If I had been there, I would have advised Tony to put her in the crush to avoid any possibility of being kicked. I had a painful lesson once, when I tried to remove a stick from a heifer's back foot out in the field while she was standing up. But Tony hardly ever asks for help. He has no farming background and started working on the farm with that admission, plus an assurance that he would do his best. This he always does, many times accomplishing impossible feats largely because he has no idea they are impossible. On one occasion, he brought two totally unrelated animals home from a far field, which I would have needed help to achieve. He has lifted ludicrously heavy objects simply because they are in the way, when I would have fetched help and a mechanical aid. Psychologically knowing that something is impossible makes it physically so and vice versa.

Meanwhile, a young man with no experience of cows was here for one day only, and asked if he could help us bring the cows and calves in from their daily walkabout. As he told us later, the youngest, Amelia – not quite one month old – 'informed' him that her mother was stuck in the pond. She had ventured close to drink and, having waded in too far, was now stuck fast in deep silt, waiting silently and patiently to be rescued. One tractor, two ropes and five people succeeded in extracting her and after a two-hour scraping, washing, brushing and drying session, with warm water to drink and mountains of hay to hide in, they both appeared to have taken the incident in their strides.

At that time, Seal, a nine-month-old heifer, was lame and taking it easy. I waited on her, taking hay, water and a grooming brush. She thanked me in words of no syllables by licking my hand and kissing my forehead. There was no obvious cause for her lameness, though I could detect slight heat in a front knee. As soon as she felt fit again, she rejoined the herd.

Filipendula needed to be milked each day, yet she declined to spend the intervening hours near the house. Although there is a much more direct route, she insisted on retracing the steps that first brought her down to the farm buildings and duly plodded up the Walnut Field, across the Humpy Dumpy, Cherry Tree and Rickety Rackety Fields to spend each day grazing by the side of her twin sister and their friends in the Lake Field, returning home on her own by the same long-distance route each evening to be milked. She behaved perfectly and displayed absolute confidence in her own decisions.

Another of the herd, Dizzie, had a large grass orme (an old word we still use for a husk) embedded tightly in the corner of her left eye. I felt an overwhelming determination to remove it for her without having to make her walk back home to be restrained in the cattle crush; it's a long walk.

I devised a plan and she watched me thinking.

I assessed the problem from several angles, talked to her, then pretended I was more interested in another cow who was standing close to her, all the while waiting to seize the right moment to grab the offending seed. My first attempt failed but, gloriously and amazingly, she didn't flee after the assault, as I would have expected; she twigged what I was up to and hung around to give me another chance. The second attempt also failed. She creased up her eye, holding the seed more firmly, but waited again. So many cows would have misunderstood my antics. Third time lucky I thought, rather tritely, but no, I missed it by a whisker: one of hers.

She was still there beside me, dear old thing. I owed it to her to succeed. I took a few deep breaths; if a fly landed or hovered near her, she would shake her head involuntarily, so each time I was having to avoid being inadvertently swiped by her horns as I dived towards her eye. I steeled myself. The first three attempts had been a bit desperate, because I was worried that each would be the only chance I got. Though she looked ready to walk off, I calmed down and took measured aim, telling myself it would be easy if I concentrated harder.

Got it!

She was so pleased. How did I know? It would take a whole book to try to explain, and if I tried, I would fail. She knew she was grateful and she knew I knew. To quote Keats out of context, 'That is all ye know on earth and all ye need to know.'

Every animal, by instinct, lives according to his nature. Thereby he lives wisely, and betters the tradition of mankind. No animal is ever tempted to deny his nature. No animal knows how to tell a lie. Every animal is honest. Every animal is true – and is, therefore, according to his nature, both beautiful and good.

> – Kenneth Grahame in conversation
> with Clayton Hamilton

CIVILISATION AND ITS DISCONTENTS

Every single person should want to be an animal rights campaigner and then no one would have to be.

The uniquely human trait of requiring proof before taking action has a lot to answer for. No animal would swallow its instinct in favour of waiting for proof. It's a handy tool of governments though, as is agreeing to hold public enquiries, which always and only delay making just reparation for avoidable tragedies. Under the guise of saving species, humans create (horrific) zoos and justify horrendous expeditions to chase, terrify, sedate, capture and control wild creatures.

When I learned that humans and chimpanzees share almost 99 per cent of their DNA, I was prompted to write this:

There is something delaying about being human
While we discuss, they suffer.
The endless, 'civilised' lengths we go to
In order to get it right

For twenty years they've worked with apes
They should know by now

They should not be subjected to the kind of things
  'they' do

They really should have freedom to decide their fate
  alone.

The last great ape would not debate, for long, upon
  the fate
Of the last 'great' man.

COMPASSIONATE SHEEP

Very late one midsummer night, during relentless rain which turned every track and path into a raging river, I fought against the current and walked up the road–river with a bundle of hay on my back and a torch in my hand, to feed the sheep. I found them blissfully happy in their little shed, in the back corner of which lay a small, tightly curled-up, fast-asleep fox cub. His coat was soaking and he had obviously been washed out of his earth and sought shelter with the sheep, who had extended a welcome to him.

There must be so many beautiful instances of co-operation in the natural world that I don't see, but it was thrilling to be able to witness this one. I hurried home and asked Rich to go back with his camera.

A few months later I saw a larger fox sitting, dog-like, outside the same sheep house, calmly conversing with two of the ewes, who were standing in the doorway. I felt certain that it was the same fox, saying thank you.

NOCTURNAL VIGILS

When you are trying to be self-sufficient, your night life can be of a rather unusual order. The other night my brain wanted to go to bed around midnight but instead I donned a coat, hat and head torch, and went slug hunting. There were hundreds of them on the leaves of the brassicas. One-and-a-half bent-double hours later, I chanced to peep at the underside of a leaf and found a vast sea of caterpillars all feverishly busy making exquisite lace leaves with tiny fretsaws. Were they perhaps not the pests I took them for? Could they have been the caterpillars of some rare butterfly trying to make a comeback? Sadly not.

But where were all the frogs? Aren't they supposed to eat slugs? Perhaps they go to bed early. Quite a few of the slugs were eating dock leaves. Have they no discerning tastebuds or are they just fine-tuning the balance of their nutritional requirements by topping off a brassica feast with a dock-leaf pudding? Or have I unwittingly stumbled on a dedicated dock-leaf-eating species, which, if 'cultivated', could be worth a fortune?

These strange musings help to stop me hating the fine, relentless, soaking rain, which never tires of proving beyond doubt that my raincoat is not even slightly rainproof. Maybe I should call my next book *Wellington Weather* . . .

Needless to say, I brought the cows with young calves into the barn out of the rain before my nocturnal vegetable-garden vigil. Writing the word 'vigil' makes me think of Virgil. I wonder what he had to say about slugs and frogs?

STRESS-FREE FOOD

The other day, just before 5.30 p.m., I took a parcel to the Post Office to be sent by special delivery. It contained ten pints of unpasteurised organic milk, half a pound of home-made butter, a carton of cream and three sirloin steaks. The parcel arrived, several hundred miles away, at 9 a.m. the following day. The recipient drank a pint of milk and decanted the rest into jugs and placed them in the fridge. She knows what her body needs. Serious illness has left her with a weakened immune system and many substances inflict damage on her. She needs stress-free food, i.e. food that has been produced on a farm where all plants and creatures are happy and are allowed to live in an unpolluted, unhurried, natural environment.

Back on the farm, as September gives way to October, the raspberries are still producing several pounds of large, delicious fruit each week and providing a paradise for the garden tiger moth who loves nothing more than raspberries and nettles – and we have plenty of both. The blackberries are wonderful this year too; even the very young calves go blackberrying, carefully picking a few ripe berries with their lips. The mulberry tree is a glory to behold, with hundreds of berries in every hue from almost white to almost black, the deep-purple berries in between a permanent reminder of the ill-fated lovers Pyramus and Thisbe, according to Ovid. The only slight impediment to free-for-all frequent plundering is the fact that it is growing in the middle of a field with two bulls grazing in it.

Does the Cold Cure Centre still exist, I wonder? I doubt it, for I believe it never achieved anything. A friend called in yesterday and she had a cold. We all stepped back three yards. Being ill is to be avoided if at all possible. Not only is it debilitating and miserable, time-wasting and inconvenient, but it places too great a burden on the other members of the family.

No more time to write: I have bread to make, cows to milk and it gets dark earlier each day.

SHEEP PLAYING

It is early April and cold, with clear skies and blossoms waiting to burst and grass growing as fast as it possibly can but not quite fast enough for our needs. We go to feed the younger sheep, the one-year-olds. They still look and behave like lambs, although once they pass the twelve-month mark they are technically called hoggets. After filling the racks with hay, we realise that fewer than half of them are there. Suddenly a cascade of wool and legs pours through the trees. It gathers speed as they hit the field, and a rush of joy swirls past us. In the middle, riding the wave of momentum created by its big companions, is the tiniest four-day-old lamb, its weeny legs a blur of speed, indistinguishable from the group but kept afloat by it; a two-year-old child snow-boarding, a six-year-old ski-jumping, an adventure cat in a backpack, running a marathon without putting a foot on the ground.

They need hay but they want to play. A flock of hungry, tired four-year-old children, getting cold and hungry but unable and unwilling to leave the glorious fun of the fair.

ON THE CHARACTER OF COWS

A town-dweller once asked me how long it took to milk a cow. She was fascinated to learn that not only does every cow take a different amount of time, even if they are producing the same amount of milk, but that they also vary greatly in the time it takes them to eat the same amount of food. What they eat also affects the taste of the milk – just ask any calf or discerning human.

All of our cows have different personalities. Some love us and respond in a gentle, accommodating way if we need to ask them to do anything. Some would prefer we ignored them. The fascinating thing is that every single one has her own particular set of likes and dislikes. If we want them to move from one field to another, some can be lured, some bribed, some persuaded, some encouraged and some can be moved by the expediency of psychological trickery, whereby we pretend that moving fields is the last thing we would wish for, and so they comply immediately.

One of the house cows, Celandine Sunshine, who is usually angelically happy to come in whenever we ask, unexpectedly decided to run off and rejoin the herd one autumnal evening. I knew I could not outrun her, so I ambled over and tried hard to make her walk in the opposite direction, ever deeper into the dark night and away from her friends. She looked at me with a degree of surprise, turned herself round, and walked home as sweetly as usual.

Cows have changes of mood. You think you can rely on individuals to behave in a certain way but they sometimes behave atypically without any warning. I often think that strong wind can affect them, and of course, a mother who has just given birth will often behave completely differently from the day before. Maybe the waxing and waning of the moon influences their mental state too; it pulls the oceans around and plays a part in migrations and crop growth. I do not underestimate the moon.

Hurry no man's cattle.

– proverb, early nineteenth century

THE POWER OF MUSIC

One October morning back in 2008, as the world's financial systems were going into meltdown, I witnessed something lovely. It was wonderful and interesting but not entirely surprising. I had opened the door to the hens' night quarters as usual, and one hen made a beeline for the barn. She walked with deliberate speed, hands behind her back, carefully lifting each foot quite high as if to protect the hem of her skirt from catching in the mud, until she arrived at the sliding doors, only to find them tightly closed. Human error, a common occurrence.

I was trailing her, metaphorical trilby pulled down over my eyes, raincoat collar turned up, nonchalantly whistling to myself each time she turned round to see if she was being followed. As I concentrated, one of the new hens pecked my foot hard. 'It's That Hen Again', I thought, my focus broken for a moment as I remembered the much-loved BBC comedy series *It's That Man Again*, which ran from 1939 to 1949 and meant a huge amount to earlier generations of my family. Both my parents quoted from it regularly.

I climbed onto the bonnet of the Land Rover, out of harm's way, and immediately gained a better view. The hen no longer noticed me. She looked at the closed doors, thought for a few seconds, retraced her steps, ducked under a gate, hurried through the barn and reached her nest site by a 360-degree detour. I have known some sheep with equal presence of mind but no cows.

I'd just reread *The Merchant of Venice* – the first time was at school in 1964, I think. Never before had I any wish to produce a play but this would translate so perfectly to reflect the monetary crisis: bankers and stockbrokers lending and gambling, people failing to repay and being dispossessed.

Relief from the financial turmoil was supplied by Lorenzo's knowledge of the calming effect of music on animals:

> *... do but note a wild and wanton herd ...*
> *If they but hear perchance a trumpet sound,*
> *Or any air of music touch their ears,*
> *You shall perceive them make a mutual stand,*
> *Their savage eyes turn'd to a modest gaze*
> *By the sweet power of music ...* *

It was only in my last year at secondary school that I had the opportunity to learn the clarinet, when another pupil decided to give back his council-owned instrument. My first attempts were a series of painful, piercing shrieks that saw the cat put his paws over his ears and my family wish to leave home. I decamped to the far end of the garden and stood playing to the horizon, only to be greatly encouraged

* Act V scene i.

to find that all the cows, on leaving the cow pen after milking, stood enrapt, listening to my squawks.

There can be no doubt that animals enjoy music. Orpheus charmed animals and birds, as well as trees and rocks, with his lyre, and Pan played his pipes. A character in *Tess of the D'Urbervilles* recounts the story of a fiddler playing in a desperate attempt to soothe the savage intentions of a charging bull, and my father played classical music on the Third Programme (the forerunner of Radio 3) to his cows every day while he milked them. He did this mainly for his own benefit, though they certainly appeared relaxed, as if they were enjoying it too.

It is nothing unusual, while taking an evening walk, to hear from a distance the milkmaid singing as she milks her cows. The cows stand patiently to be milked, and give every expression of pleasure upon hearing the song.

A milkmaid who had a large number of cows to milk once told me that some songs met the approval of the animals, while others were received with disapprobation.

'There is Cherry,' she said in her pretty patois, 'if I was to sing the "Men of Harlech" to her, she'd kick. She do like a love song best. Iss indeed! And if I do sing "Jenny Jonse", she'll rub her head 'gainst my arm. Then there's Lily, well, she do like something uncommon lively, the "Fair Maids of Merioneth" especially. As for Lovely, why, if I do drawl out "Poor Mary Ann" she'll stand like a Briton. But there'd be a rumpus if I was to begin "Hob y deri dando!" They'd get nighty all of a sudden, and never quiet down until I'd sing their favourite songs.'

– Marie Trevelyan, from *Glimpses of
Welsh Life and Character*

AN EXTREMELY WINDY NIGHT

One morning in late January we woke up to discover that our main cattle yard – historically called a 'cow palace' but known to us as the cow pen – had collapsed overnight, bringing down a power line in the process. The building had been erected shortly after the First World War and used by previous owners as a poultry house and a traditional cowshed. For many years, the farm supplied the milk for many of the people in Broadway, and we still have one of the original, embossed milk bottles, retrieved intact from the stream many years ago. The building had an attractive shape, with a pointed roof not unlike a Cotswold house and a window in the gable end above the entrance, giving it an almost face-like appearance. But it was constructed with timber beams that had already started to rot, and poor-quality precast-concrete walls, already in shabby condition when we came here.

According to the morning news, the storm had affected the whole country. A number of people had been killed, and thousands of properties damaged. We discovered later in the day that several trees on the farm had also blown down, some damaging fences. Fortunately, the cattle had access to an outside yard with feed racks and a water trough and must all have fled outside as the cow pen started to come down. But now we suddenly had an enormous clearing-up job to do, in the middle of winter. The task fell entirely on Richard's shoulders as I could rarely leave Mum for more than an hour at a time. He had to find a way of housing nearly fifty cattle on a farm whose few other buildings were already filled with cattle or hay.

Rich disconnected the power lines and bedded the homeless cattle down temporarily in their open yard; luckily no rain was forecast. He telephoned various neighbours to see if he could rent any space, but there was nothing available, and in any case he quickly abandoned the idea altogether when he remembered the implications it could bring for our TB-free herd.

Instead, he moved all the bales from two bays of the hay barn with our loading tractor and then went to Evesham to buy a huge tarpaulin to protect them from rain. He dismantled the water trough and pipe from the yard and installed them in the barn, enabling him to rehouse the six cows with the youngest calves. They gave it their seal of approval, possibly because they were able to reach over the gates to get extra hay from the bales in the next bay. But this still left the stronger cattle to spend another night outside and this time rain was forecast.

My mother, who was a born farmer, even though seldom a physically active one, always tried to use her brain to help us solve problems. She suggested putting the stronger cattle in our conifer wood. In places the trees were still growing close together and this provided a good windbreak. There was a small stream, and a stoned area where we could put feed racks and a proper water trough, though we would need to buy a hundred metres of pipe to make a temporary connection. It took Rich two days to get everything ready, so the cows had to spend another night where they were.

The day their new woodland home was ready also happened to be my mother's eightieth birthday. I drove her in the Range Rover in front of the cattle in order to show them the way as Rich encouraged them down the drive, through the poplars and into the conifer wood, where he had rolled out big bales of straw among the trees and filled the feed racks with hay. The cattle were hugely relieved to be let out of their cramped confines. Some started eating straight away, while others went exploring or rubbed their heads on the trees. The arrangement proved to be very practical. Although we ended up using more straw than usual and had to buy in extra, the cattle were very happy and comfortable.

AT THE BIDDING OF BULLS

The first signs of snow at Kite's Nest split the camp, with one group of cattle coming home and one group staying out. Dizzie and her youngest daughter, Dizzie Lizzie, took up residence in a special pen and within minutes they were both 'smiling' from ear to ear. The older Dizzie had always been more than capable of standing on her own four feet, but the legs above those feet grew a bit arthritic with the passing of the years, and if there was competition for hay she would move aside to avoid being jostled rather than hold her ground as she used to. In the safe haven of the pen she could have ad lib hay undisturbed.

The cattle who elect to tough it out are given hay in the shelter of a large belt of woodland. They appear not to notice, let alone mind, the snowflakes that look so magical in the headlights and feel so wet in the dark.

One of the two bulls stayed out and one came home. Red Bull is a 'homing bull'. He has his own paddock, which he loves, opposite the house. Every so often we gather a group of his wives, former wives, mistresses and hangers-on and take them all to a lovely area of the farm with the freedom of half a dozen fields. He always appears to be content for the first few days but then he gets bored and comes home; his reappearances are unannounced and can be a touch alarming when he soft-footedly sidles up behind me in the dark just as I am switching off the lights after milking the house cows. The gold bull, Peter Goldfinch, stays with the herd unless specifically invited to come home. I once received an email from my friend Mary Cooper, then ninety-six:

It was the '40s . . . far away in the Golden Valley on the borders of Wales.

We had two oil lamps for outside work, that's all and you had to manage. One night, milking in the dark – because Bert had the lantern up his end of the cow shed – I felt the warm nozzle of the bull Bill on the back of my neck as I sat against the cow on my milking stool. Bill's chain must have come off from round his neck. Fortunately I had the sense to sit still against the cow on my milking stool – and I called to Bert . . .

All was well in the end . . . no one was particularly concerned . . . but I've never forgotten the feel of that warm wet nose on the back of my neck!

If I see Red Bull grazing not very far away, I might well throw him a greeting. He will hear but will not stop grazing nor acknowledge me. If he comes as close as he can, though, to attract my attention and ask for hay, he requires not merely an acknowledgement but an undertaking. I usually deliver on time, though he is happy to wait if I am genuinely tied up with a job. If I forget altogether, I can be sure he won't. Cows will usually ask and ask until they receive. Red Bull puts in a request and then carries on as before, confident that his bidding will be done. I have been known to remember only after going to bed, and then deliver the promised hay wearing pyjamas.

WILDLIFE CORRIDORS

Britain's largest nature reserve could and should be the odd-shaped, snaking snippets of land that border our extensive road network. Glance at parts of any road atlas and you would be forgiven for thinking this island consisted entirely of roads. These thousands of acres should be safe havens for every species that aggressive farming methods banish. Beyond the wall or fence or hedge, wildlife corridors should also be beyond the control of the tidying culture which, when enacted with a flail mower, spells homelessness and reproductive quietus for every insect, mammal, nestling, pupa or seed case drawn, as they invariably are, into the vortex of the flail mower's making and pulverised to oblivion.

With their promise of irresistible destinations, Britain's roads still beckon, but the cost of petrol has shrunk the mileage's appeal and I attempt to show how much travelling I can do on the spot.

The wildflowers on this farm determine a pattern of frequent visits: oxlip, wood sorrel, early purple orchid, half-hidden violets, marsh marigold, yellow flag iris, free-climbing honeysuckle, agrimony and ragged robin. One unmowable meadow of head-high grasses had escaped our detailed attentions, however, until a seeming thousand-mile search one July for a seemingly lost twin calf saw me criss-crossing the entire paradise with its astonishment of butterflies clouding my passage and its soft carpets of equisetum brushing my knees.

My absolute determination to find the missing calf meant that not one square yard would be unsearched. My fruitless quest, while mother and calf were quietly reuniting themselves two fields farther off, became more and more of an exquisite compensation as I stumbled on flowers hitherto unknown to me. So many species happily squashed side by side and so many minuscule habitats effortlessly sliding in and out of one another: dry, high bits, home to gorse-nesting goldfinch, clumps of tough sedge and meadow floor of squelching, seething, beetle-rich flowering rush, meadowsweet and delicate, purplish umbellifers.

What could be better than the knowledge that our four-month-old heifer twins are so independent that we sometimes don't see them for days on end? We slip and slide in the farm vehicle to deliver hay to the members of the herd who prefer not to come down to the barn and the twins are not there . . . nor there . . . nor anywhere that we can safely search without becoming stuck in the mud. So we return home disappointed but, having spoken at length to their mother, not even slightly worried.

Every evening we eagerly anticipated the dusk-driven display of several thousand corvids right above our house. We used to hear the jackdaws first, then see them, swimming in shoals towards and almost into one another, and then came the rooks fresh from their parliamentary debates, synchronising with, complementing and perhaps competing with their smaller cousins, hurling themselves up and up and down and across, laughing at their own nimbleness until there was hardly candle power enough to 'light their dusky way to bed'. Sadly, these spellbinding sights are now a long-distant memory, partly thanks to the dramatic decline in insect populations, and partly because rooks are not widely enough recognised as the beneficial birds they are, but seen instead as agricultural pests. They do not eat significant amounts of grain but they do feed on species such as wireworm and leatherjackets which themselves destroy crops.

The hens also 'delight and hurt not'. They pay us (in eggs) and entertain us, and two sit on the doormat waiting to be carried home.

THE HEART OF THE MEADOW

It is impossible to photograph a wildflower meadow, to try to capture it and show it to others like a trophy. It is the magic of the sound and movement, the explosive jump of the grasshopper – 'green little vaulter in the sunny grass' as Leigh Hunt called him; even the bumble bee caught in the patient spider's web is part of the wildflower meadow story. It is the surprise of the delicate grasses, so tall and fine you can't see them till they brush your face, and the endless variation in height that gives the depth, from the ground-hugging, starry-eyed tormentil and its cuddle-carpet of bright, tight new growth, through the mid-height nearly black, gone-to-seed heads of the rush, to the bleached-gold foxtail and tight balls of down that are the seed heads of all the hawkbit family. It's the pounding heat on the quiet greenness, the grand, sparkling betony and the surprising yellow of dyer's greenweed. You cannot photograph the anticipation and excitement of seeing the hazy brilliance of the whole thing waving slowly; each stem and flutter independent and exquisite components of the whole horizontal mural. The meadow is a composite creature, whose heart you can hear only when you become part of it too. When you sit, eye-high with the grasses and in the flight path of dragonflies and chimney sweeper moths, time stands perfectly still until you stand up again, just as childhood is regained as long as you stay in the sea, no matter how old you become on land.

MILKING BY BRAILLE

I found myself the other night milking a cow with my left hand while sending a text message with my right (I really think there should be a law against that). The cow had only one quarter of milk to spare, her calf having demolished the other three, which is why I had one hand free. The reddish sunset that had been illuminating the milking parlour in an almost surreal way had vanished and I was milking by Braille, which was straightforward enough. But sending a touch-text was almost beyond me; there was no way I could see to proofread it as my reading glasses were in the house.

As the cream rises to the top of a bottle of unhomogenised milk, if you are lucky enough to know where to buy some, so it seems that the cow keeps the cream at the top of her udder, and you get cream only if you milk out a quarter completely. If any of you remember stories of your great-grandparents nipping out into the field to snatch just enough milk for a couple of cups of tea from any cow willing to stand still for a few minutes, you can be pretty sure it would have resembled shop-bought semi-skimmed.

When hand-milking a cow, it is a good idea to rest your forehead on her flank, the soft piece of flesh that arches over the udder and joins her hind leg to her abdomen; this provides the dairy-person's early warning system. Even the sweetest-natured cow can move a foot inadvertently at the wrong moment. A non-inadvertent move might send both milker and pail reeling. But the cow's decision to move, or kick, has to be communicated from her brain to her leg, and a firmly resting forehead feels the message just before the leg gets the order to move, usually giving enough warning to whip the bucket out of harm's way.

Of course, it's also a good idea not to get too comfortable, nor lean too heavily, as the rhythm begins to dull your sense of self-preservation. She moves a little, you lean a little to maintain your proximity. She moves a little further and, bucket between your knees and both hands occupied, you fall off your stool.

Some cows like to be talked to, but some require a respectful silence, whereupon communicating with the forehead assumes a greater importance.

Let the Cows Speak

If you want milk, you can have,
Not with pleasure but without resentment,
Everything we can spare; calves come first
    That's fair.

If they are large and need it all
    That's that.

If they are small then you will get
    Quite a lot.

But beware. They will grow and you will know
    When they need your share.

We will come, as our forebears came
Down the immemorial dusty lane
And you will give us in return
Apples and sweet hay and a grooming.

We will stand, chained, content
While you sit, doubled-up and bent
And gently draw your diminishing three quarters

But you will find one day, although we came
    Just the same
That our udders have been quite
    Sucked dry.

You will still have milk for your tea
    For your tea
But your bread will go unbuttered.

Your lean times must be averaged
As your accountant averages your income
And overall you'll do well

And we will live, with allowances for
Human interference, as we wish to live.
    Before we die,
With human interference.

                                – Rosamund Young

PSYCHOLOGY AND SLEEP,
DITTO AND SHEEP

It is often said that if we get less than eight hours sleep a day, we are not going to be performing optimally, and the less sleep we get, the worse we will be. Knowing this has a psychological as well as a physical effect. Quite how different I would feel if I experienced only the physical effect I shall never know.

Sheep can be psychologically affected, though I very much doubt if they ever worry about getting too little sleep, or too much. Our sheep know me well, but in the back of their minds they are always wondering if my friendliness is the prelude to some restraining activity such as foot trimming. They do not have to think like this when it comes to other people, and they can like or dislike them purely on merit. I am pretty sure they like me; in fact I believe I can (almost) prove it, as once when I was unable to see them for eleven weeks they positively mobbed me on my return, looks of disbelief in their eyes, for all the world as if they had concluded I had died, but were keen to show me how glad they were to be wrong. I was moved. I admit to being slightly jealous when they give their affection to 'other people', while still viewing my loving overtures with suspicion for weeks and sometimes months after I have wooed and grabbed them, for their own good of course.

Time after time our sheep show what immaculate character judges they are; they never get it wrong. We learn from them. The sheep like all children, little ones, unless they are zooming around in an unfathomable whirr. Tiny people approach and the sheep stand still – the same sheep that wander off, grazing single-mindedly, when I want to get a closer look at them. They sense the absence of ulterior motive. They do approach quite a few adult humans too, often walking right past me in the process. I catch them, trick them, tip them up and inspect their feet; my friends, walking companions, and casual visitors do not. Sheep remember, for years and years. It would take a huge event to make a sheep wish to harm or even threaten a person, though it has been known. Mrs Dangerous, for example, knows how to threaten: create eye contact, mock head-butt, letting you know what's in store if you don't keep your distance. But it's only to protect her privacy, and the spaces she creates for her lambs. She won't 'come looking for you'.

If I walk into the deep, dark wood, in winter when the light has gone, I know that the cows will alert me to the presence of strangers or dogs. The sheep rarely spend any time in the wood, but they would do the same.

DINK

On 'ninth night' in 2009 I was out in bitter cold and serious darkness, on my way to visit a cow. Although there was a waxing moon, she did not throw her beams into the corners of the cattle yard.

Earlier that day, Dink, then nearly eight years old, had slipped on the icy concrete and 'gone down'. One vet and three strong men managed to magic-carpet her into a warm pen and after a day of devoted attention it was my pleasure to do the night shift. We gave all our blankets away many years ago to our solicitor's wife, who was on her way to help save oil-drenched sea birds, thus Dink was given a duvet. Once she felt thoroughly warm, she discarded it.

Dink was a very brave cow and exceptionally sweet-natured. She trusted us so totally that she was not even slightly worried that she could not stand. The vet could not discover what injury she had sustained, and she showed no hint of pain as he carefully investigated. We watched and nursed and groomed and spoiled her and enjoyed getting to know her three-month-old daughter rather better than we might otherwise have done. She ate like a horse (though I bet cows eat more than horses!) and drank with a grateful 'smile' as we held a bucket in front of her.

We all realised that she might never recover but over the ensuing days she made quite a few new friends. Men who might have fed her as one of the herd, and cast a stockman's eye in her direction, now became properly acquainted with her on their visits at all times of day and, quite simply, grew to like her a lot.

The true but unsung heroes of the countryside are the variously termed horse-slaughterers or knackermen. They do the sort of work that most people prefer not to think about, when an animal is struck by a lorry in the middle of the night or slips on ice. In our sanitised world, we should remember the huge debt we owe them.

Dink had a great life, a happy, optimistic injury time and the humane death she deserved at the hands of a skilled and compassionate professional. She suspected nothing, eating apples from my brother's familiar hands. After she was put down, a post-mortem revealed that she had broken her pelvis when she fell and could never have recovered.

Dink's young daughter, pleasingly but also rather alarmingly, didn't appear to miss her at all. Having been taught to take milk from a bottle and teat and having enjoyed the extra attention she had been given, she now took her place in the herd like a miniature grown-up with all the *savoir vivre* she needed.

COLD FEET

Cows seem fairly philosophical about snow. If they have water, food and shelter they can forgo their daily walking and grazing routine and settle down quite happily. Not so the hens. The first time they encountered snow they looked almost offended; five soon got the hang of it but one played off-ground tag for all she was worth, hating the idea of getting her feet cold.

The sheep positively enjoy themselves. In the winter months it often isn't until after midnight before I take them their apples and hay, partly because the boring task of water-thawing takes so long. They come skipping out of their little barn to greet me like adventurous children, and as I have to open the gate to their paddock in order to carry in the five-gallon container of water, they take the opportunity to escape for a bit of extra-curricular playtime. For the humans, not falling over on the icy mud is the main lesson to be learned, and a hard one it is!

EXPLOITATIVE ANIMALS

Diary entry, 15 March 2009

Spring has arrived. All day and to our extreme delight, the thrush has been rinsing and wringing the ear, as Gerard Manley Hopkins put it. The first squeak of the dawn chorus began at 5.30 a.m. and for once my microphone was ready.

As I sit here at the kitchen table, facing due north, the jackdaws that had gathered in the ash and sycamore trees on the top of the hill have just lifted off as one, and are at this moment careering over the house; it is 6.28 p.m. The cattle that stayed out last night are very glad they did, and those who came in were champing at the bit this morning to go back to the fields.

Three-quarters of an hour ago two cows and a totally unrelated six-month-old bull calf came home: Dorothy, Charlotte and Mr Dis. Charlotte, who has the loudest voice of any cow in England, was deputed to 'ring the doorbell' and I obediently trotted some three hundred yards to open the door (gate) for them. There is a footpath running through the farmyard; just as the trio emerged from the field, two walkers came past and the two cows stood magnificently and patiently, with Mr Dis just behind them, like two priceless cars waiting at a zebra crossing, waiting for them to pass.

Red sky tonight but more in the south than the west. For the next few hours individuals will wander nonchalantly in, and I will be at their beck and call. Someone should start a Society for the Protection of Humans from Exploitative Animals.

Better a good cow than a cow of a good kind.

– proverb, early twentieth century

HOUSEBUILDING

Hand-made, my foot. Hand thrown, handicraft, hands-on. The blackbird's nest, balancing safely near the edge of the just-above-eye-level shelf in the not-much-used pump house, was made entirely without hands. Unbelievable. He sings into the almost dark and she sits pretty. She knows there will be rain and she knows she will be dry.

They're all at it: rooks and wrens, marsh tits and dunnocks, dashing about and working wonders. Meanwhile, two mallards stroll round the plots with newly granted planning permission, to see which one will be nearest to the school of choice for their chicks. The hens find the wind fierce and cold, and huddle on the doormat. I close one of the two doors to the porch and inadvertently shut one hen out. She has been busy somewhere, and when she turns up five minutes later, not surprisingly she can't open the door. This small individual, possessed, presumably, of a 'bird brain', uses it to astonishing effect and turns herself into a temporary hummingbird. She leaps three feet on to a half-inch-wide window ledge, flapping her wings at the speed of light to stay impossibly there, and pulls faces at me as I wash the dishes, forcing me to run to the door with profuse apologies.

How many new houses do we need each year? We have an army of builders hard at work already: long-tailed tits, goldcrests, wrens, a pigeon on top of a tall pile of old Soil Association magazines, something incredibly architectural in a niche in a redbrick wall, chaffinches, coal tits, bumble bees, ants, who may not know that they work full-time for the green woodpeckers . . .

Shakespeare tells us about bees, near the beginning of *Henry V*: 'They have a king and officers of sorts . . .' And John Clare's similar observations on ants capture what most casual walkers might so easily miss:

> *What wonder strikes the curious while he views*
> *The black ants' city . . .*
> *Such government and thought . . .*
> *And what's more wonderful, when big loads foil*
> *One ant or two . . .*
> *A swarm flocks round to help . . .*
> *Surely they speak a language whisperingly*
> *. . . they have kings and laws . . .*

WATER

There's a lot to learn about sheep. From as far back as I can remember, I have listened to farmers talking about how amazed they are that sheep never seem to drink much water. I too am amazed. I watch them constantly and it's hard to decide why they seem to behave a bit more like camels than the cattle I am so familiar with. Some farmers even went as far as to believe and state that sheep never drink water at all. Even at a very young age I knew this could not possibly be true; water has to be handy at all times for whenever any of the sheep need it.

It's quite common for a sheep to drink up to two gallons after giving birth, but if water is not available, they won't go to seek it until their lambs are strong enough to accompany them. No ewe would walk off and leave newly born lambs unattended. Having said that, even tiny lambs will sip water. Water is a stress-buster and cows and sheep will calm down and stand still for their offspring to suckle once they have drunk sufficiently. In fact, it is pointless to try to restrain a ewe or cow to try to teach its young to suckle unless it has been offered water first.

When we bought our flock of northern rare breeds, mostly Shetlands, all 114 of them went a whole week without drinking. I had expected them to drink copiously to help them recover from their long journey here.

Milk is made from a recipe of grass and water but don't quote me exactly. Lactating animals will drink far more in order to make milk. The quality of milk from cows eating hay or silage is very different to the milk they produce from grass. It will be the same for sheep and deer and rabbits and every mammal.

Sheep self-medicate and often eat what we humans believe are poisonous plants. Out foraging, I occasionally taste unfamiliar plants and usually wish I hadn't. Animals know without stopping to think precisely how many leaves and shoots and blades they need of each plant. We have to learn, either by trial and error or by trusting someone else's knowledge. We simply do not consume so many potentially valuable additions to our diet because we find it easier and more convenient to buy everything, even though plants are all around us if we look.

There are so many things to learn about sheep: the number of possible (and impossible) lambing positions, foot-trimming requirements, optimum length of grass, frequency of moves to fresh pasture, complementary grazing species and the likelihood of ewes getting cast (stuck on their backs, which can be fatal), to name but a few. Grass is the most appropriate food for ruminants, but it too needs to be as natural and unstressed as possible, since the quality of food an animal eats affects the quality of the produce it gives us.

Grass covers a huge proportion of the surface of the world, and forms the background to the rhythms of life.

The rain-soaked fields create a percussive squelch under your feet and you can hear and feel a rhythmic harmony as you walk, picked up by the metronomic tap of your zipper pull, your own heavy breathing as you fight the gusty wind and the squeak of your arms against your mud-splattered coat. Any and every pop song could begin from such an introduction. Dry grass crackles but late spring soft grass waves silently, allowing the insects to hum the tunes.

DUPED BY A COW

A lifetime's experience gives me the confidence to assess when a cow is preparing to calve. One evening several days ago, one of the Dizzie family started displaying 'classic behaviour'. She left the herd and took up temporary residence in a far-flung field, looking a bit uncomfortable and with her tail slightly raised. I went over, had a quiet word and made arrangements to visit her again in an hour or so. When I returned, she was in the same spot but lying down.

Much as I would have preferred to leave her where she was, to calve in a nice, clean field (forever mindful of a surgeon's comments when doing a hip-replacement operation, that it would be safer to perform his surgery in the car park than the operating theatre), I instinctively felt that as she had made no progress, she would need help. I asked her to accompany me home and she complied with extreme good grace. I thought she must realise I was intending to be of use. I was wrong.

Sometime later, having happily consumed several halved apples and a flake of hay, she told me politely but insistently that if I had finished with her, she would like to go 'home'. I tried hard to dissuade her; I even tried to ignore her. This proved impossible.

I opened the gate and off she went, calmly and purposefully into the dark. I followed. I had no need to lurk, however; unlike many wild and some domestic creatures, she had no intention of leading me on a wild-goose chase to prevent me finding her 'nesting site'. She was treating me all along as a friend and confidante and simply waiting for me to come to my senses.

We crossed several continents (or perhaps a few hundred uphill yards), and as she arrived at the top of the hill where I had first found her, a handsome, pale-grey bull calf bounded out of the woodland, popped over the low wall and tucked in for a good drink of milk. There were no recriminations; she bore no grudge, he had not been anxious at her absence. I think and hope that he had been sleeping off a previous milk binge-drinking session.

Perfection. I could give myself the sack.

The ox knoweth his owner . . .

> – Isaiah 1:3

ROUGH-SHOULDERED BY A HEN

One May Day, with the may blossom already out, a hen laid a beautiful egg in the kitchen. She had asked politely enough to begin with, but when I explained that I'd just washed and polished the kitchen floor . . . and it was a lovely day . . . birds singing, sun shining and, no, on the whole, no, she had to stay outside, she rough-shouldered the tiny-bit-open kitchen door and barged in. She had made up her mind that it was the ideal place to lay, and woe betide anyone who tried to stop her.

Her elected laying site brought back memories of *Beyond the Fringe* in the 1960s with Peter Cook, Dudley Moore, Alan Bennett and Jonathan Miller, and Cook and Moore's subsequent play (on words), *Behind the Fridge*.

RUMINATING ON RUMINANTS
ANTHROPOMORPHICALLY

Dot, having calved without fuss or ceremony at the civil-ised time of 8 p.m., did not initially appear to like her new daughter and mooed at her loudly. The calf, hardly fifteen minutes old, blarted back. I'd never seen or heard anything like it. By 7 o'clock this morning they liked each other a lot and are now happily ensconced on a greensward carpeted with buttercups.

Later that night, in the darkishness, Gold Giselle waited patiently by the barn gate for someone to open it. She had got that far because the field gate had been left open so that the house cows could return straight to their grazing, as they had requested. Two hours earlier, Giselle had appeared serenely happy, standing with her son, admiring the view (I hope) and watching the sun setting. The first labour pains evidently included warning signals and home she came for human intervention. When I woke up, an exquisite pale-gold female calf was lying by her side in the paddock. The calf had been trying to enter the world upside-down, and Richard had spent the night helping her to be born the right way up.

Her first day was one of perfect weather bursting with birdsong, and at 2 p.m. she skipped and bounced and looked a bit pleased with herself.

I am constantly surprised by what our cows teach us. If they like someone, they can display a huge degree of tolerance bordering on the indulgent. Recently our herd has been introduced to a new man called Will, who is young, enthusiastic, kind, capable and speedy. To my astonishment, the cows truly indulge his predisposition for speed. Previously they have ruled me with a benevolent rod of flexible but unbreakable willpower; they require that I walk at their pace and wait patiently each time they stop for a nibble of dock seeds (evidently delicious), a mouthful of walnut leaves, or anything else that might take their fancy. Their new friend walks them in at twice the speed of sound and my fears that they would object by giving less milk have proved groundless; they give more, happily, and I look on and learn.

A week ago I observed what appeared to be proof that an eight-year-old cow could recognise and respond to the voice of her three-year-old daughter, out of sight eight hundred yards away, with many fields and trees in between and after six months of separation in different groups on the farm.

I was thrilled, amazed and moved by the mother's request that I open a gate, and by the subsequent sight of the two of them lying back to back, rubbing heads, grazing side by side and, no doubt, communicating in other ways beyond my ken. The mother's most recent calf, Carline III, ten months old and only recently weaned, was left behind with the half of the herd that Carline had chosen to leave.

Carline II needed her mother and Carline III could manage beautifully without her for a while. I should not have been surprised. I have spent my life watching freshly calved cows responding instantly to the voices of their infants and, equally often, the infants reacting to and obeying without hesitation the whole range of maternal communication from a whisper to a bellow. So why should a grown-up daughter's voice not be familiar?

Some parents educate their children at home for a few years and thus hope they will be better prepared for school life later; some children never go to school. One of our young heifers had educational theories of her own. She hid her new-born daughter from human and other bovine eyes for a whole week. I tried hard to find the calf but was reassured by seeing the mother's udder at times full and at times empty.

THE FIRE

It was mid-December. Rich had been up all night dealing with a difficult calving and was just about to collapse into bed at 7 a.m. when the lights went out. He dragged himself back downstairs assuming he would need to mend a fuse, and found the kitchen, which he had left just a teeth-brushing session earlier, filled with smoke. He had carefully checked the wood-fired central-heating boiler as usual, but what we didn't know (until the Fire Service investigators told us) was that recent work to line its chimney was fatally flawed. The builder had estimated that a wooden beam would go into the wall only six to nine inches. In fact it went in twenty-three inches, and the steel flue, installed from the other side of the wall, was just an inch away from the wood. The heat of the metal caused the beam to char back under the sitting-room floor and eventually catch fire. The settee directly above burned quickly, along with the electricity wires behind it.

It was bitterly cold, but we had to wrap Mum in as many clothes as she could wear and leave the house. We sat in the Range Rover, with the engine and heater on, watching in a state of stunned resignation as six fire engines arrived and we faced what we feared would be the whole house disappearing in front of us.

Suddenly remembering Mum's medicines and essential food, I went in through the back door to collect as much as I could carry. The firemen saw me and ushered me out of the front with worried faces. I returned again and again, unable to envisage being able to keep Mum alive without the things on which she relied so absolutely. I felt an unreal sense of achievement for managing to collect so much, mixed with desperate fear. The firemen were hurling our possessions out of windows and doors, saturating everything, and yet I was managing to climb the back stairs as if nothing was happening. Quite soon it became impossible and I retreated to the vehicle to watch.

The sitting room was completely burned and two other rooms were badly damaged but the firemen saved our house. Although we lost many treasured possessions, we could never feel anything other than gratitude to be alive.

We had let our cottage to three men who were working locally on a big temporary project and they had all gone home for an early Christmas break. Richard contacted them to explain what had happened. They each insisted that we take immediate possession of the cottage and accepted his offer to find them alternative accommodation. We moved into the cottage and Mum and I remained there for five months. As soon as Rich had got the power back on, he moved back into the house to start making it habitable again.

A combination of shock and disorientation left Mum very unsure where we were. The word 'cottage' gave her a mental image of the cottage she lived in as a child. The stairs were very steep and I couldn't leave her alone at all. Richard had an overwhelming workload and needed whatever help I could offer, so I developed a routine of fetching and milking the house cows after Mum had gone to bed.

Some time later, we were all back in the house at long last, and a skilled and cheerful carpenter called Dave was making and mending. Mum offered him a cup of tea and asked if he had ever read or seen a Shakespeare play. 'No, but I would love to,' he told her. 'My wife likes me to sit and watch televised adaptations of Jane Austen and I hate it.'

Mum asked me to fetch three copies of *King Lear* (we own at least six).

Mum and Dave and I read *King Lear* for half an hour that day and every subsequent day till Dave had finished his job here. He understood a lot and asked Mum to explain anything he didn't. Richard would have loved to have joined in but as Mum needed me physically to prepare multiple tiny meals each day, and mentally for every reason anyone could think of, Rich had almost all the farm work to do and missed out on the literary treats.

When Dave left, we were not quite halfway through the play. A year later I got an email from him, addressed to Mum: 'You changed my life. I enrolled at evening classes to do English Literature and the whole world has opened up to me; I cannot thank you enough.'

There is no wealth but life.

– John Ruskin, from *Unto This Last*

I WOULDN'T THANK YOU FOR
AN OFFICE JOB

A spider, small and pale with thin, almost translucent legs and a dull buff blob for a body, was hurrying across the floor the other day when he noticed my foot. He stopped and thought for a minute, then tiptoed off in the opposite direction.

When we were still milking at Kite's Nest, there was a time when we had three house cows: Philadelphia, Filipendula and Laura. All three had their own calves, but if they produced more milk than the calves could drink then we would have the surplus.

One day when I went to fetch them, they were reluctant to come as they had been given access to the aftermath grazing – the regrowth after haymaking – and had been captivated with pleasure at the delicious taste of the grass. Consequently, they had no desire whatsoever to leave their high pasture.

Eventually I managed to persuade them, but instead of walking the three hundred yards a crow would have flown, they made me (and their calves) follow 'over hill and over dale, thorough bush and thorough brier',* even though it was getting dark and pouring with rain.

---

* *A Midsummer Night's Dream*, Act II scene i.

I loved every minute of it.

No, honestly . . .

On and on we went, along precipices, up mountains, across streams.

I've never enjoyed myself so much.

After all, I was losing weight incidentally, getting fit and learning the farm's topography by Braille. I got only a few thistles in my knees and it took only a few minutes to get out my glasses and a magnifying glass and a torch and a pair of tweezers each time.

I wouldn't thank you for an office job. When I got home, Rich milked the three of them so that I could recover. They had made a bargain with me that they would be allowed to return to the herd straight after being milked, so Rich duly opened the gate.

He found them all, a little while later, lying snugly in the hay barn, unable to stop themselves from 'smiling' at their own brilliance in doubling back when he wasn't looking and thus avoiding a night of rain.

I read that the Roman senator Cato was also a farmer. In *De agri cultura*, he listed what he considered to be the four wisest uses of agricultural land, as follows:

Profitable cattle raising
Moderately profitable cattle raising
Very unprofitable cattle raising
To plough the land.

Cato also had an eye for practical detail and agricultural psychology. He advised that you need 'barred feed-racks. The bars should be a foot apart; if you make them so, the oxen will not toss their fodder out.' And he wrote a memorandum for farm managers: 'Have special care taken of the oxen and be a little indulgent to the oxherds so that they are readier to take care of the oxen.'

ON THE PHYSIOLOGY OF WOOL

I know a bit about knitting but nothing about spinning. Luckily, however, I know someone who does. She is a one-off, a true self-taught genius who spins wool that feels like silk. She took the wool from our four sheep in individual, named bags: Angelina, Tealeaf, Sapphire and Lydia. The feedback she gave us, knowing nothing of their upbringing, was fascinating and revealing. In short, the sheep with the best fleece by a mile is the one that was reared on her mother's milk and never had the slightest setback in her entire life. Angelina's fleece is 'like cashmere' and almost every scrap is usable.

Her twin sister, Tealeaf, had to be reared on a bottle because their mother had only one teat that produced milk. Tealeaf is a fine specimen with a good brain and good powers of reasoning and memory. The deprivation she suffered before anyone realised she was getting no milk manifested itself in the quality of her wool. It is short, too short to be spun, and poor in every respect.

The other two sheep have wool of good length and generally fine condition, reflecting exactly their respective, though minor, traumas in early life.

Time and again I have noticed illness, injury and misfortune resulting in physical imperfection (however slight). Brain power, on the other hand, does not seem to be affected; in fact the reverse is often true. Having been at a physical disadvantage, these animals had to use their brains more, to survive in a group where size and physical strength generally equate to priority.

Tealeaf's story was initially a very emotional one. It was a difficult time here, with Mum being particularly unwell. A farmer we knew slightly rang to ask if we would adopt a ewe and her two lambs as they had almost no grass. He failed to mention that the ewe had only one functioning teat. He and his wife delivered the trio on a day when Mum was too unwell for me to leave the house, so I spoke to them out of a window and pointed to the paddock where they should leave the newcomers. I wanted to get to know the arrivals, but it was impossible for me to leave Mum until the next day, and that involved a large measure of deception.

Mum fell asleep and I sprinted down to the paddock to meet the new sheep. The ewe looked fine and one lamb was beautiful but the other one was thin and hunched, clearly starving. The sensible course of action and one Mum would have insisted on, if I'd told her, would have been to euthanise it. I felt an overpowering wish to save the lamb but I was taking on a huge task.

Initially, after pretending I was going to brush my teeth, I speedily heated some milk and ran a hundred yards to offer it to the lamb. She resolutely refused to drink, instead going to the ewe and apparently suckling. For a moment I felt my diagnosis must be wrong. On my third hasty visit she drank. She and I both heaved a sigh of relief. She got fatter and I got fitter but it was a hard secret to keep. Once she had some strength she tried to steal from the other sheep. She had no success, but in the process she earned herself the name Tealeaf: Cockney rhyming slang for thief.

A NATIONAL ASSET OF
IMMEASURABLE VALUE

I read that an editorial in the *Eastern Daily Press* on 17 April 1946 stated:

> Four hundred and seventy-three men . . . are waiting to hire smallholdings from the Norfolk County Council . . . there was never greater need than there is now to keep enterprising men of this kind inside British agriculture . . .

And here is G. A. Squires, writing in the book *The Small Farmer*:

> The inherent capacity of the English country labourer for making the most of a self-occupied small acreage of land ought to be encouraged in every possible way; it is a national asset of immeasurable value. What is more, the independent small cultivator of this type is potentially very much more efficient than the large farmer . . . the best internal security for our future lies not in a Bank Balance but in a Fertility Balance.

I go to the bookcase and re-read William Barnes's Eclogue, 'The 'Lotments':

> *I'd keep myzelf from parish, I'd be bound,*
> *If I could get a little patch o'ground.*

I'm not the one who has the sheep; it's the sheep who have me.

– Axel Lindén, from *On Sheep:*
Diary of a Swedish Shepherd

THE BOVINE SCARLET PIMPERNEL

We once had a cow (I shall keep her anonymous for fear of notoriety) who realised that if she physically stopped the Range Rover she would get more apples than her actual allocation, as we tried to divert her attention while we made a bid for freedom by dashing to the farthest cow rather than the next nearest. It was Mum's greatest pleasure to give apples to the cows, cut in half of course, however out-of-season and expensive they were. But this sweet bovine was undeterred and instead of following us, she would watch calmly until we slowed down and then take the most efficient shortcut. She would either stand four-square in front of the vehicle to prevent us from moving or she'd trot alongside poking (good-naturedly) at the front wheel with her horn and bashing (gently, so as not to cause a dent) at any part of the bodywork she could reach. At this point, we'd stop, get out, stroke her and tell her how clever she was, give her several halved apples and disappear in a puff of . . . well, yes, diesel fumes.

COMMUNICATION

There is something amazing about the ways cows can communicate with each other. One cow may have such presence that she always goes through each gateway first without having to assert her superiority. One might employ a nod or a wink and another might need to do the dignified bovine equivalent of an 'after me, if you don't mind'. And then of course there are the many layers of subtle conversation that sometimes produce the desired effect from quite a distance. As A. B. Tinsley wrote in *Horse and Cart Days*, 'There is always an order of seniority [among cows] which is rigidly adhered to, as can be seen when a herd passes through a gateway . . . the old bull . . . didn't give a damn whether he came through first or last.' It is not on the whale scale, but impressive nonetheless. And once they have run out of sign language, there is always the spoken word . . .

All animals are individuals. Some cows are timid but the majority can hold their own in the herd. And, just like us, some enjoy being stroked, groomed, praised and appreciated and some like to be left alone. There are endless variations on the theme of greeting. Some people shake hands; some always embrace. Some of us like to wave, bow or otherwise acknowledge one another in a non-tactile way.

Humans often strive to be diplomatic; cows and sheep are simply direct.

Even in the past, most people would not come into direct contact with sheep. Flocks would have been tended by their shepherd and sheep were only seen from a distance or in passing, frequently in such vast numbers that detecting individual traits would appear an impossible and perhaps unthinkable feat. Cows are bigger, slower, and less inclined to swarm off into the distance. A shepherd's life was lonely, isolated and unremitting but he (and as far as I can tell it was almost always he until only recently) would have known a great many members of his flock as individuals and would have relied on some of them to lead the others and actually help him in his daily tasks. In *The Shepherd's Life*, James Rebanks says he knew he could trust one of his older ewes to lead the flock to a sheltered place during a snowstorm that deterred most of them from moving at all.

Sheep need to take their time to get to know people too. As they were so often preyed upon in the wild, their instincts would have been tuned to mistrusting people. In the relatively short time we have kept a big flock of sheep, there have been several notable occasions when a sheep of one year or four or ten has decided to befriend me; I suddenly find it standing patiently by my side, having appeared from nowhere, and the trusting bond is sealed. I could decide to befriend a sheep but it wouldn't work. They are the ones who decide who to trust.

Cows, in my experience, rarely moo for help when

trying to calve, though they make interesting and decodable noises. Sometimes when a sheep baas, it is so obvious that it is asking for help that it really does seem to be speaking. The best thing of all is the unfailing ability of all our cows and sheep to character-judge human visitors. We learn from them, gratefully.

Cows are my passion. What I have ever sighed for
has been to retreat to a Swiss farm, and live entirely
surrounded by cows – and china . . .

– Mrs Skewton, in Charles Dickens,

Dombey and Son

A FIELD OF FROGS

I would love to say I saw a field of frogs the other day . . .

On the basis of what I actually did see, I believe many statisticians would have allowed me to assume that the field was indeed hopping. The cows and sheep were in the meadow and the grass was long. Each time I stopped to speak to a particular animal, the grass moved at my feet and a frog hopped on his way. I made the assumption then, and I still believe now, that the field was full of frogs. Beautiful creatures, frogs are environmental barometers, like butterflies. I saw a garden full of those recently too: commas, peacocks, emperors, painted ladies and for a fleeting moment I felt sure I was looking at a large tortoiseshell, notwithstanding the fact that they are extinct, officially at least. Frogs were here long before someone built this house on their traditional 'marching route'. So now they often wait by the door for us to allow freedom of passage to the back garden.

A BALANCED DIET

Whilst listening to Radio 4 one morning I heard about a clean-air survey in which everyone was being invited to participate, with lichens the focus of attention. I had always believed that lichens indicate pure air, but now I discovered that some types thrive on pollution and that experts can tell much about air quality from close observation.

Nature has everything if one knows where and how to look and as long as we humans don't tidy, poison, cut down or otherwise destroy. Herbal remedies are there for the asking, though once we have forgotten our instincts, who do we ask? The cows know, and the sheep, the ants and the beetles know. Wild and domesticated animals nibble and browse plants as they find them. People tend to chop, brew, stew, dry, powder and alter the herbs, often extracting the essence of a plant and producing a substance far removed from its original form: simultaneously far more powerful and more dangerous. The sheep actually seem to take greater risks with self-medication, though I expect that's just my human perspective. They nibble what you would expect them to nibble: leaves of every bush and tree they can reach, thistles, grasses and weeds of every description. But I have also seen them consume ragwort and cuckoo pint (Arum maculatum), with its startling red berries that I was always warned not to touch.

It's all very well writing about how delightful our sheep and cows are and how brilliant and how companionable our hens, but the main reason they are so clever and so happy is because they eat organic food. They breathe pure air and drink pure water and eat what nature intended them to eat. The ruminants graze grass and they love it.

Worldwide, millions of acres are devoted to producing cereals to feed to ruminants; this makes them human competitors. Vast acreages of grass are needed to lock up carbon, and ruminants grazing the grass produce healthy meat. People cannot eat grass, but they can eat cereals, and healthy protein from pasture-fed animals is a valuable addition to a balanced diet. According to the *Ten Years for Agroecology in Europe* report published in 2018 by the Institute for Sustainable Development and International Relations (or IDDRI, to give it its native French acronym), 'If we stop feeding grain to livestock we can feed Europe organically.'

We have watched our farm animals getting cleverer and happier since 1953, while also monitoring the effect of the food we eat on our own health and the health of our friends and customers.

In August 2009, my mother injured her leg, sustaining a deep wound. She was small and frail, having never been well for some forty years, and she had, by most standards, an extremely limited diet. But everything she could eat was totally organic.

The wonderful nurses who came almost daily to dress her wound – and there were at least nine of them plus two doctors – all independently remarked that the speed of her recovery *must* have been due to the quality of the food she was eating.

So, there you have it, NHS: give all your patients (and staff) organic food and you're home and dry. Of course, eating 'conventional' food won't kill you; in fact it will almost certainly sustain you long enough for you to live to regret eating it.

A CRITICAL THREAT

Our late friend Joyce Smith was a member of the Women's Land Army during the last world war and a talented language teacher who suffered from chronic fatigue syndrome (often referred to as ME), a very debilitating and depressing illness. We were full of admiration for her, as she was unfailingly cheerful and determined to get well. She once emailed us, saying:

> The first time I went down with this was in the 1960s, just a few weeks after a holiday in S. Italy. DDT was being widely used, on government orders, to eradicate the malarial mosquito, so I daresay the water supply, etc., was contaminated. I believe there are around one million ME sufferers in the USA and around 100,000 in the UK. Perhaps one day food produced using pesticides and herbicides will be properly priced at the checkout, covering the cost of the damage it does to water supplies, soil, human health, etc. Then only the very rich will be able to afford it and everyone else will have to eat organic.

DDT, an insecticide with devastating impact on the environment and human health, has been banned for agricultural use since 2004, but still, GM crops and our interference with nature for commercial gain pose such a critical threat that I can no longer indulge the huge pleasure of writing only about all the living things that creep and pad, bluster, float, flutter, scurry, bounce, dash, plod, jump, fly, climb, crawl, dig, wriggle or merely grow on this farm.

Let us be in no doubt: large companies with the expertise to produce genetically modified crops are not doing so out of altruism. They have no wish to feed the world to prevent people starving. They wish to provide the crops to feed the world to make money. Also, in my humble and very concerned opinion, anyone with the skill and (probably) brilliance to succeed in genetically modifying plants is likely to be clever enough to wish to avoid ever actually eating any of the results of their labours.

If genetically engineered organisms are released into the environment, they can never be recalled. There are already many instances of unpredicted consequences with scientists fighting nature rather than trying to work in harmony with it. If one day in the future GM technology stands accused of endangering the health of the planet, it will be too late to do anything about it.

The World's Need

So many gods, so many creeds,
So many paths that wind and wind,
While just the art of being kind
Is all the sad world needs.

– Ella Wheeler Wilcox

INTENSIVE CARE

One Wednesday in late October 2009, Mum was still sleeping in her bedroom at the top of the house she loved, though I feared she wouldn't be able to manage that for much longer. It took her a long time to climb the forty-four steps up to bed and even longer to come down in the morning but she cherished the achievement.

When Mum got to her sitting room each morning, she had two bells she could ring, one for me and one for Rich. The frailer she became, the easier it was for her to forget whose turn it was and on this particular day she got Rich, unaware that he'd been working till 4.30 a.m. on two successive days trying to finish writing a book chapter. He got up and looked after her, until my return to the house allowed him to go back to bed. Despite being tired he felt fine, but as he was getting undressed he suddenly got a violent pain in his stomach. He could barely move. He managed to get halfway down the back stairs and shouted for me to call an ambulance, something he'd never done before. It was two hours before the ambulance arrived (unknown to me, if I'd said he had pains in his chest, the call would have been given a higher priority). Initial examinations in hospital failed to reveal a cause. They tried to send him home but he knew something was very wrong and refused to leave.

More than thirty hours later, at 9.30 p.m. the following day, a young registrar who had just finished operating was sent to see him, seemingly for experience, since no one else had a clue what was causing the problem. He was very open and told Rich that the general belief was that he was making a fuss about nothing. He did, however, ask slightly more detailed questions and suggest a chest scan.

By 11.30 p.m. the registrar had identified the problem. He said Rich had an aortic dissection. He drew a diagram of the aorta and explained that the inner lining had split and that blood was escaping into Rich's body cavity. He said it was quite unusual and he'd previously only read about such things in textbooks. Rich went from being a nuisance to a high-priority patient in less than a minute. He was given a range of drugs and fluids intravenously to bring his blood pressure down and rehydrate him, and was connected up to various monitors.

The planned operation had to be repeatedly postponed. On the Monday morning I received a phone call from the hospital. Rich was seriously unwell and the operation wasn't without risks. I got the message and was determined to go down, but it would take some organising. I stayed up most of the night preparing as much as I could of Mum's meals for the next day. I learned later that one of his closest friends who had also visited that morning had been told he only had a fifty–fifty chance of pulling through.

Late in the evening I rang the intensive care ward, and was told that Rich had survived the operation though he wasn't entirely out of the woods yet. He was heavily sedated with morphine, but things looked hopeful.

The day came when he was due to be discharged. Marisa, one of his lovely colleagues from the Soil Association, kindly offered to bring him back. It was wonderful to see him home, but he was still in a lot of pain and on strong painkillers, with a bottle of morphine to sip when it was unbearable. He went to bed and stayed there for the next week. Initially my workload increased further as I had him to look after as well.

HOW NOT TO WORRY

Could you find your way to a specific location in Africa without a map? A swallow can. All birds are brilliant: swallows, buzzards, kestrels, robins, wrens and hens. Everyone who can should keep hens. Not only to love and enjoy them, and reap the immense health benefits of perfect eggs, but to respect and learn from them: how to live, how to make the most of each day and how not to worry.

You ask if they were happy. This is not a characteristic of a European. To be contented, that's for the cows.

– Diana Vreeland on Coco Chanel

EGALITÉ, FRATERNITÉ

Maharan, our oldest ram, doesn't seem to like sheep but he really loves the cattle and spends as much time as possible in their company. We are not sure whether he thinks their friendship is worth winning and therefore sees them as a challenge, but he definitely sets the rules. Maybe word got out when he had to teach the bull a lesson, after which none of the herd ever pushed their luck with him. All winter they get along famously. He eats hay at the rack with them and they always let him choose where he wants to stand before they start.

Maharan teaching Prometheus a lesson was a sight worth seeing and we are grateful we were there to witness it. One winter we took hay in the Land Rover to the cattle and sheep, and were just about to come home when we saw the bull and the ram, one at least ten times the size of the other, start to eat the same flake of hay. I have no doubt, knowing him as well as I do, that Maharan would have shared the meal very happily. Prometheus wanted it all and pushed the ram away. Maharan recovered his balance and composure and walked back to share the hay. The bull pushed him away again, this time with more force. Maharan had no choice but to reverse and charge at Prometheus, crashing heads with him. The bull retreated in defeat. The ram continued eating the hay.

UNDER THE STARS

As winter sets in on the farm, several calves take to staying out with a vengeance. Long after darkness has dropped down, a quick head count reveals a discrepancy, and the identities of the offending youngsters are ascertained. Some mothers go pretty mad, mooing for their calves, and we have to let them back out to plod up the hill to find their children. But others – notably the members of the Dorothea family, who wouldn't worry if you paid them to – just stay silent and eat their way into the night. Around 3 or 4 a.m., sometimes, maybe, a quiet, sweet, questioning moo asks why the miscreants are not tucked up with their peers. On cold and unforgiving nights, I wear my wellies out traipsing the landscape, fearing that perhaps, this time, the 'baby' will be lost and hungry and in need of guiding home. No such thing! Always and always she (and I've only just realised this, it is inevitably 'she' and not 'he' who stays out) is totally happy, under the stars, by an oak or in the depths of a hawthorn hedge, agreeing to come home only to please me.

One little heifer calf squeezed through a fence in order to play–graze with some friends. Consequently, she arrived home at dusk by a different route from her mother and ended up in the wrong barn. Neither she nor her mother seemed to notice until 5 a.m. when they both decided to moo very loudly, until some human servant crawled out of bed to reunite them. The next day she did the same thing and arrived home walking just behind her friend. She instinctively followed the other calf towards one barn then stopped and thought for a moment, turned herself round, and trotted off in the opposite direction. It really looked as if she was thinking, 'I'm not going to make the same mistake again.'

Allegri's *Miserere* was long kept as the exclusive possession of the Sistine Chapel but one day the young Mozart heard it there, went home and wrote it down, note perfect. I need an artistic version of Mozart here, at night, after dark. The things that happen here 'after hours' cannot be captured by a camera. They need a genius to observe, wonder and recreate. What this mythical human would see is really beyond my powers of description.

Come the winter solstice, a few individuals choose to stay at home, relentlessly eating hay and silage. Others venture out to see what old – in fact, ancient – varieties of grasses are still nibbleable despite the frost.

They begin to come home in ones and twos and threes at teatime and as they would all like to have a pen to themselves, allocation becomes the issue of the day. I am reminded always of a television programme I saw many, many years ago. Someone was interviewing (or more accurately just talking to) a group of boys who had spent their whole lives in care. One eleven-year-old, when asked what he would most like, replied, without any self-pity or expectation of ever having his wish fulfilled, 'I would just adore to have a bedroom of my own.'

One snowy evening, Cocoa III and son managed to grab one of the prized 'bedrooms'. It's a room with a view but no water on tap so, last thing at night, I have to carry buckets to replenish what's been drunk since teatime. It was very late, cold, dark and slippery, so to save time I left the door open in between journeys to the tap. Little Mr Coco, who had appeared to be sleeping soundly, noticed the open door and hopped out for a skip in the snow. He was a bit surprised that his feet went in a different direction from his nose but he persevered and I had a pretty tricky time persuading him to return home. A little earlier I saw that the two bulls (three-quarters of a ton Red Bull and half a ton Peter Goldfinch) pricked up their ears as soon as my penknife starting slicing apples in half for the two cows I milk by hand and came over to look pitiable. An impossible task, but their amusing attempt earned them the desired apples.

Most of the cattle choose to go out even when the snow is too deep for any grass to be found, but for the past few days the slight amount of snow has not hidden the taller, tougher, older grasses and the short-lived winter sunshine has tempted many of the cows up the hill to graze. This one-day story started as usual with the great gate opening, but old Carline and her daughter stayed behind. I walked towards them to encourage them out, and as Carline moved forward, I could see she was limping. They spent the day eating hay and I kept waiting for the ideal opportunity to inspect her foot. Carline is one of our tallest, biggest, heaviest cows and inspecting the hind foot of even a smallish calf is not easy.

Very late, long after the whole herd had come home and everyone was lying contentedly and drowsily, I paid Carline a quiet visit. She looked very comfortable, half sprawling, half lying, and I hoped she would not feel inclined to move. I stroked her neck and worked my way down her leg to her foot. She twitched a bit. Even from a recumbent position, cows can pack a powerful kick but Carline was really comfortable and I think she could tell I was trying to be helpful. I managed to stroke my way down her leg to her foot by leaning over her flank, and gingerly began to probe between her clays to see if there was an offending object causing her to limp. Then I realised someone was breathing down my neck. Carline's eldest daughter, four years old and well on the way to being as large as her mother, had come extremely close to check whether I might be hurting her. She watched my every move; her proximity restricted my right shoulder. I wriggled a bit in order to carry on with my investigations but she just moved in even closer. I felt that trying to persuade her to back off might have made her suspicious that I had harmful intentions; by permitting her to watch, she would be able to see I was doing good. I then realised I was also being watched over my left shoulder. Carline III, sixteen months old now and less assertive than her sister, had overcome her shyness to come to her mother's aid. I massaged old Carline's ankle with a copious quantity of comfrey ointment, gave her love and thanked her for trusting me and then burst out laughing as I noticed her newest baby, one month old, quietly watching everything but without the least concern; she had learned to trust me on the day she was born.

Some cows are just cows. Well, yes, in the sense that some people are just people. All the people you don't personally know: the ones you see on television queuing for the Boxing Day sales, the crowds at football matches, the audiences you hear only when sitcoms are recorded, and of course all the people in other countries who you have never seen and will never see. Yes, cows are just cows in that sense, but once you get to know one, or many, you know that they are all individuals with incredibly varied personalities.

I take my last, late walk round to check everyone is all right. I see calves in motionless, sleeping scrums, oblivious to their mothers' whereabouts. I see family groups like the Dorothy family with nine close relations all within a few feet of one another. I spy Red Nell who really has love written all over her face as she lies with her daughter tucked under her chin. Now that's what no camera could capture.

DROVERS AND ROVERS

It's official: our sheep know a green Land Rover from a red one. Even so, they prefer travelling in a Range Rover if given the choice.

NATURE AND NURTURE

Who'd have thought that an inclination for extortion could be inherited? Many years ago we had a man named Tony Griffiths helping us. With no previous experience of farms or cattle, he showed an eagerness to learn to milk and, most crucially, an understanding and respectful affection for the cows. He refused to learn their names but developed the skill of identifying each individual by searching for odd markings, often as unnoticeable as a tiny spot of white on the tummy of a black cow. I would frequently see him bending down to check who was who. Although he liked the cows, initially he didn't believe they could have individual foibles.

Felicity, white and large, elegant and very good-natured, had a calf that died at birth in the middle of a run of ten calves which thrived. Tony milked Felicity by machine and on day three, he was excited to tell us that she had found a subtle way of forcing him to give her more food, by raising her hind leg in veritable slow motion and then swiftly returning it to the ground when fed. He knew perfectly well that she intended to kick the bucket over . . . but never did.

Nigh on thirty years later, I witnessed Felicity's great-great-granddaughter behave in an identical fashion.

The Cow

The friendly cow all red and white,
I love with all my heart,
She gives me cream with all her might,
to eat with apple tart.

– Robert Louis Stevenson

A NEW FIELD OF STUDY

I love the English language. A new field of study could be pursued in a library, laboratory, warehouse or even an actual field! The two cows we are hand-milking at the moment behave completely differently towards Richard and me. They call the shots, but they take different liberties with each of us. My new field of study moves from cow pen to barn to field and back again. When it is my turn to milk, I might well mention to Rich what one or other of the cows did and he often expresses surprise bordering on disbelief; one cow might be compliant and gentle with one of us and difficult with the other.

The relationships between cows and their daughters are a beautiful, sometimes oddly complicated, rich and fascinating field of study, but the human–animal trust test involves a very 'grown-up' give and take. Fairness is paramount – they must be rewarded if, for instance, they agree to come in when they would prefer to stay out – but they play their parts with integrity if we behave according to their rules. They bully us a bit, they plead but they don't cheat. The sheep walk towards some people and actually show excitement and pleasure, but this is by no means universal; you are either one of the chosen or you are not. Freedom of association is something that people value; animals do too.

DON'T FORGET TO PUT THE MOTH OUT

That has a Pythonesque ring to it, but liberating moths that have been accidentally trapped in the house must rank among the most important of daily activities.

HERDING AND DROVING

A study lasting three years, at zero expense to the taxpayer, has come close to proving conclusively that one particular cow has decoded the human psyche. If she is part of a herd being moved from A to B for a bona-fide reason she is totally content to move. If she is being moved for a dubious reason, however, or being brought in to some sort of confinement she senses this, no matter how hard the human perpetrators try to hide their motives (which reminds me of *The Midwich Cuckoos*). She remains friendly, unafraid, calm and compliant on the surface but after walking a few hundred yards, she notices a lapse of human concentration and speeds off into the sunset, or deep dark woodland, whichever is the nearer.

Our Lleyn sheep are fairly easy to drive and shepherd but the Shetlands seem always to appoint an outrider or two who are on permanent alert, ready to signal to the others the perfect moment to break ranks and race incredibly fast, uphill if possible, to outwit and exhaust us. We call them 'leaders' and they have stripey faces and disruptive ambitions.

FIT FOR A KING

. . . ash wet or ash dry
A king shall warm his slippers by.

– Lady Celia Congreve,
'The Firewood Poem'

The most elegant and useful tree, the ash; no other will yield up sticks to light a fire when new, green, damp or wet. We have many ash trees, at all stages from saplings to giants, and some of the most magnificent are out of sight in the wood, having grown taller than they might otherwise have needed to because of the competition from faster-growing conifers. One remarkable specimen is so tall, and on such a steep incline, that one has to get on all fours to worship it. There is also an unusually tall ash growing on a steep rise of fairly poor soil, out in the open field. It has already lost several massive limbs and one day it will topple over. When it finally falls, some branches will crash and crack into pieces that will fly and bounce and roll downhill and some large bits will stay where they are. And the cows will eat the leaves and crunch on the twigs and rub and push and play, and bit by bit, over the ensuing years it will all come home.

EXUBERANCE

Many of our cows come to 'have a chat' on my evening visits to the herd. More often than not, however, the youngest calves decide that it's playtime. There is no doubt that they can see in the dark as accurately as they can in the daylight and, having enjoyed their first week of life tucked up under their mothers' chins, they are now ready to launch into independent nightlife. One will stand up and stretch, look back over its shoulder to check that its mother is contentedly dozing and trot off to call for a friend. This duo might then meet up on the 'street corner' with two titches from the next field and excitedly skip and gallop to wherever it is 'at' this particular night. We stand and witness games of pure joy, celebrating speed and grace and dexterity, as they dash and zigzag between and around the darkening shapes of older siblings, aunts, cousins, ex-babysitters et al.

Dot looked as if she would calve three days ago but she still can't be bothered, so I kept watch, tactfully. Yesterday she had a daughter – and she certainly didn't need me. By the time I arrived, maybe half an hour after the birth, the calf was washed and 'dressed' and ready for school. Who needs people anyway?

Three things are not to be trusted: a cow's horn, a dog's tooth, and a horse's hoof.

– proverb, *c.* thirteenth century

A RETROSPECTIVE DIARY ENTRY

9 May 2011

There are frantic times and busy times and, yes, a few quiet evenings and long summer afternoons when we sit and talk on the lawn that the sheep saved us having to mow. And there are lonely times too: times of illness, injury and nursing others. I have spent more time nursing than being nursed, forty years more, but at the time of writing I had just spent four unplanned days in a Bristol hospital.

Will, the highly capable young man who had taken over my job of milking our two cows once a day to maintain my mother's demanding dietary needs at the time when she needed my brother or me constantly present, always communicated with me by text. As a rule, he could solve most problems and accomplish everything solo but one day he sent me a text: 'I can't get Filipendula in. She is being pig-headed. Please can you come to help?'

Moments before, I had succeeded in sending my mother to sleep with self-taught reflexology. I slipped out of her room, put on my wellingtons and ran towards where I knew Will would be. Feeling invincible and with my ash staff in my hand, I leaped the fence, sprinted a bit and produced the commanding words and gesticulations needed to coax Filipendula in. It had taken only a couple of minutes and I was keen to return before I was missed.

I relaxed, and as Will turned to close the gate, I laid down my stick and approached the cow to stroke and gently admonish her. In my rush, I made the cardinal mistake of putting myself in a corner. She charged at me, and for the only time in fifty-eight years of living on a farm, I had not left myself an escape route. With no time to think, I tried to climb the high wooden fence, but she plunged her misshapen horn into the back of my knee and ripped the flesh. I continued climbing. She did it again. I screamed to Will and he raced over and pushed her away as I sank to the ground on the other side and felt my wellington start to fill with blood. I peeped and saw some bone exposed, quickly hiding it with the skirt of my dress.

'Are you okay?' asked Will (who hadn't seen what had happened).

'Yes.'

'Are you sure?'

'I could do with some water.'

He disappeared into the parlour, shouting back that he could only find a bucket.

'That will be fine.'

He found and filled a four-pint jug with cold water and I drank every drop.

'Are you sure you're all right?' Will asked again.

'I'm fine. Milk the cow.'

I was clearly in shock. If I sat still, I thought, it would go away.

Will finished milking the cow and let her go back out into the field, took another look at me and marched towards the house, saying, 'I'm going to fetch Rich.'

I can go too, I thought. I took the belt off my dress, tied it round my thigh and sloshed after him.

The paramedics had to cut my (new) wellington off to get at the wound and it took them over an hour to stabilise my leg with an inflatable device so they could take me out to the ambulance. I spent two days in Frenchay Hospital in Bristol, waiting to be sewn together again (being some way down the list of priority operations) and two days recovering sufficiently to be sent home – as soon as I had learned how to go up and down stairs on crutches.

A kind neighbour offered to collect me and drive me home. I found Rich, having had no sleep at all for four days, bearded and dishevelled, and Mum, fortunately, not quite as sharply aware of events as she would have been even a few weeks earlier, very happy to see me but not over-worried.

Rich had been doing the night shifts for some time, ever since we needed to give our mother round-the-clock care. Luckily, he has always been nocturnal, but doing day shifts as well pushed him to the limit. His arthritic knee, awaiting surgery, required industrial quantities of strong painkillers, but he refused to have the operation while Mum needed help because he'd been warned that the condition was so serious it might require amputation. It took another three weeks before I was fit to work again, and two months before I no longer felt afraid of every single cow.

Two friends insisted on coming to help us during this difficult period. Matthew Greenhalf, a gifted young landscape painter who helped Rich look after Mum, and Anton Wylam, at the time our talented computer fixer, held the fort in many ways, including cooking meals for us all. As we learned later, Anton was already dying from lung cancer and in considerable pain, though he had not told anyone. He only just made it to sixty. The abbey in his hometown of Tewkesbury was completely full for his funeral. We clearly weren't the only ones to have been helped by him.

A few weeks afterwards, while a friend was sitting talking to Mum, Rich and I were able to go to the cattle crush to look at a lame cow. We were both on crutches and negotiating the rough farm track carefully, so we did not at first notice Jamie and Nick walking behind us, leaning on makeshift walking-stick crutches and swinging their legs in perfect imitation.

During the six weeks when I wasn't allowed to drive, Jamie drove Mum and me round the farm, but she was thrilled when I could resume our daily outings, something she had always looked forward to and enjoyed. She was getting weaker and frequently nodded off as soon as she had settled into her seat, tired with the exertion of getting there and happy to have done so. The minute we got back home she would wake up and ask to go again. She always insisted on going back to the kitchen for a cup of tea first and since the effort of negotiating the path and steps tired her out, she usually fell asleep again at the start of the second trip.

LIFE AND LUCK

Mummy died on 24 July 2011 at 4.39 a.m. How the end came is a long, sad and interesting story but not one for this book. Richard and I were each holding one of her hands and had been for many hours as she slipped away, able to speak and then only to nod if she needed a sip of water, then in and out of a strange sleep. It was 4 p.m. before a doctor finally came to certify her death. We then went to finish baling and carting our last field of wildflower hay. It took half an hour to find the back-door key; one of us, usually me, had been in constant attendance for years and all that time we had never locked the door, just bolting it at bedtime.

Eighteen years earlier, on the day my father died, we also had hay to make. They would both have approved: farming was what they lived for. They are buried on the farm. Rich had insisted on digging Dad's grave on his own but was far too weak to dig Mum's. The ground was rock hard. Jamie and Nick volunteered to do it for us and took great pride in making a neat job, even lying down inside to make sure it was long enough. Mum had enjoyed their company on many occasions, and they both liked her and wanted to be involved.

Richard had been pushing himself too hard, and not many days after Mum died he collapsed and had to go to bed, where he remained for six weeks with his strength only slowly returning. My forty years as a carer had not prepared me for my new role: running the farm on my own, making decisions (my first) while also ministering to Rich. I was out of touch, too. When I needed new wellingtons to replace the ones that had been cut open by the paramedics I jumped in the car and drove to buy a pair, only to find that the shop I remembered had moved fifteen years before. I sat in the car park and stared at the branch of Next that stood where Countrywide had been, before finally walking in and buying two pairs of leggings and a striped dress.

During this time Rich, from what I feared might well be his deathbed, issued various pieces of advice, which sibling affection and diplomacy made me feel I should follow. One such was a recommendation that I place an advertisement in the classified section of our local newspaper to try and find a tenant for the spare room, to help pay the bills. The combination of Rich's knee injury and Mum needing constant care had necessitated employing extra help on the farm over many years, which had significantly increased our borrowing. I composed a brief advertisement and quickly found myself arranging to interview five single men.

The last of the five interviews was with Gareth, following an unusual email which I later discovered had been written by his concerned friend Pete, keen to help him escape his claustrophobic lodgings. He arrived at the appointed time, I showed him round and Gareth said he would like to live here. I felt he was the only candidate with whom Rich and

I could share our house. In the end it took four months before he moved in. Being a perfectionist, he asked if Pete, who does not drive, could redecorate the bedroom to his taste, and thus I saw him for one minute every evening when he collected Pete, having dropped him off unseen each morning on his way to work.

One evening I invited them both to stay to supper. Rich was up and about now. After the meal, Gareth asked if he might come to watch me milk the house cow. I had elicited only a small amount of information from this rather secretive man, but had learned that he had studied botany at university and had then changed tack and qualified as a cabinet-maker. It flashed through my head that I would have to take care he was neither hurt nor frightened by the cows.

Annoyingly, the milking machine, which usually worked perfectly (having been bought second-hand in 1953), refused to produce sufficient vacuum and the teat cups dropped off repeatedly. No one had watched me milk for years and I was struggling not to swear out loud. After watching silently for some time, Gareth very quietly asked if I would like him to have a go. To say that I was alarmed would be an understatement.

'No!' I almost shouted. 'I'll run to fetch Rich.'

I wanted to make him come with me, to prevent any possible mishap, but he didn't move, so I raced to the house. By the time Rich had put his boots on and walked the forty yards to the cow pen, Gareth had mended the machine and milked the cow.

I was incredulous. Apparently, he had spent every spare second since he was about ten helping on a neighbour's dairy farm in North Wales. He began offering to help me on the farm each weekend and I started looking forward to his visits more and more. We worked hard but we had fun. One day I decided to continue a conversation we'd had by sending him an email. Twelve days later I received a reply. It was very worth waiting for, though the wait had been torture. Richard helpfully said to me, 'Never underestimate the power of the unsaid.'

The curfew tolls the knell of passing day,
  The lowing herd wind slowly o'er the lea,
The plowman homeward plods his weary way,
  And leaves the world to darkness and to me.

– Thomas Gray, from 'Elegy Written
    in a Country Churchyard'

THE OLD COPPICE

It must be at least sixty years and probably a great deal more since any coppicing was carried out on this farm. Some time ago I fought my way through a jungle of very aggressive nettles, slid down a steep bank, jumped a stream and clambered to an almost inaccessible bit of woodland to reappraise the old coppice. Sweet chestnut is coppiced in some parts of the country and makes very fine fencing stakes, as the ratio of heartwood to outer wood means it is more resistant to rotting than most other woods. We only have two specimens: one alive and one long dead.

In our little piece of old woodland near the house I can see the remains of coppiced hazel and ash, though in a different part of the farm there are magnificent willows, which might be able to be brought back into the rotation. I'd love to live in the wood, learning the craft by trial and error, my mobile left at home and just circling, mewing buzzards overhead and primroses at my feet for company.

In *The Shepherd's Calendar* John Clare talks of the company of primroses:

> *Where woodman that in wild seclusion dwells...*
> *Crushes wi hasty feet full many a bud*
> *Of early primrose yet if timely spied...*

The sight will cheer his solitery hour
And urge his feet to stride and save the flower . . .

I am fired with zeal to make a hurdle and then to make dozens. Not to mention bean canes, fencing stakes, and rails, though for those I'd have to learn to cleave. I've been absorbed in videos of men, often in pouring rain, drilling holes in old logs, hammering in hurdle uprights, weaving whip-like hazel branches and twisting and coaxing them round and round to strengthen the structure every six inches. I'm wondering if it's too late now to restart coppicing; maybe the branches are so strong they would not want to regenerate if I cut them at an angle of twenty degrees. We shall see. I won't be able to resist cutting a few, even though coppicing should take place when the trees are dormant. For hazel, it would take only seven years to see if I've succeeded, and a mere twelve to fifteen for ash. I should be able to weave a few hurdles while I'm waiting for the regrowth!

By the piles of feathers I find, the broken snail shells and empty eggs, I conclude that quite a few creatures are murdered in the wood here every night. The killing goes on in daylight too, with kestrels treading the air, waiting to pounce, buzzards unzipping rabbits, and the evidence the owl pellets provide would keep any detective busy. When we came here in 1980, the woodland was dense and dark, with no pathways. Every dusk we would hear screams which we thought and hoped were either foxes or deer, though they sounded for all the world like someone meeting their end.

John Clare heard such sounds, too. Again from *The Shepherd's Calendar*:

> *The badgers shrieks can hardly stifle fear*
> *They list the noise from woodlands dark recess*
> *Like helpless shrieking woman in distress*

He goes on to describe in detail the much talked-of alleged murder of a maid by her illicit lover after dark, which put fear into the hearts of those who heard screams and shrieks at night.

ANOTHER USE FOR HAZEL WANDS

We've got nine million hazel bushes (okay, if you don't believe me, you count them).

Anyway, the point is that even though we have nine million hazel bushes, we never get a single nut as we've also got nine million squirrels – approximately.

It would be fair if they waited till the starting whistle went, as I could give any squirrel a good run for its money, but they have already started long before the nuts are ripe, and that is just plain cheating. I'm going to choose the hazel wands not bearing nuts (I mean, I don't want a single squirrel to go to bed hungry) and try my hand at constructing a little shelter, just big enough for a seven-year-old friend.

After a few days the tiny house takes shape. It is a bender, a type of dwelling that woodland workers have lived in for centuries – circular, domed, quite strong, but a bit more than slightly imperfect. Many good lessons have been learned though. The more uniform the poles the better, and if nicely tapering they twist together admirably and hardly need string (making string from nettles is on the agenda).

WHAT IS THERE TO BE GLAD ABOUT IN NOVEMBER?

Here's a brief list:

Ever increasing circles of toadstools in the open pastures and fascinating clusters of fungi in the woods; hundreds of broad helleborine in the dappled shade of the poplars; hedgerows groaning with haws; charms of magical goldfinches swaying on the thistledown and flocks of flitting yellowhammers; a huge moon on the horizon; berries on the holly; sloes on the blackthorn; and playing circuses of jackdaws and rooks, sky-diving, hang-gliding, free-wheeling and generally living it up in the evening skies for as long as they possibly can.

ON BOTTLE-REARING LAMBS

Shepherds have many tricks up their sleeves but in the past, before the days of bottled milk or powdered ewe-milk substitute, there were not many options when a lamb's mother had either rejected it or died. A shepherd would have had to catch a ewe with excess milk and try to think of a way to persuade her to allow the unwanted or orphan lamb to suckle. If the ewe's own lamb had died, its skin would have been removed and tied on to the hungry lamb in an attempt to trick the ewe into accepting it as her own. These days, with both colostrum and milk substitutes widely available, the task is definitely easier. We intervene and rear a lamb on a bottle if, for instance, a mother of triplets knows she doesn't have sufficient milk for three and determinedly prevents one of them from suckling.

There are unwritten rules to remember: regular intervals between feeds are essential, it is better to underfeed rather than overfeed and immeasurably better to have the milk too cool than too hot. Lambs love warmth and they love motion, so if you need to move one in a vehicle it will enjoy the ride. If I only have one 'orphan' to rear I usually take it with me everywhere in the Land Rover. I know it's safe and it has company and will go to sleep when I need to get out to work.

I don't just want such lambs (all lambs actually) to be healthy and grow fast, I want them to be happy; happy lambs grow fastest. It only takes a few minutes to wrap one in an old jumper and cuddle it. As soon as it's warm, it will relax and be far easier to teach to suckle from a rubber teat.

So often during lambing we are too busy to stop for meals, especially if the weather is difficult and/or wet. Richard hands us flasks and we float through each day on rain and tea.

A WOODLAND ARRIVAL

Philadelphia IV disappeared into the wood one autumn evening, as many a heifer has done before her, and found a charming spot under some towering red oaks to calve. The wood covers sixty acres and I chose to start looking for her at the top; she chose to calve at the bottom. I was a bit concerned, not to say worried, that she might be needing help, but it was such a beautiful day that I couldn't but enjoy myself. My purposeful meanderings took me past beech and hawthorn, Douglas fir and redwood cedar, turkey oak, ash, maple, hemlock fir (I think) and multitudinous fascinating and dangerous-looking fungi.

Philadelphia had already licked her new son dry and fluffy and he had suckled and settled down to sleep. It would have worn her calf's little legs out to be route-marched back to rejoin the herd. She was close to the stream, so I carried hay to her and over the following three days she fed him, and he slept and grew. By the time he was strong enough to walk (or run) a marathon, they had become so accustomed to their woodland existence, they didn't wish to leave it. The calf, understandably, believed he was the only calf in the world. Whenever I was a bit tardy carrying her daily hay, Philadelphia would walk home and remind me, but the calf didn't come with her.

One day she clearly told me that there was something amiss: not much, but something I could help her with. She finished her hay, then walked back towards the wood. Jamie offered to accompany her and soon saw that the calf was on the wrong side of a fence and could see his mother but not reach her. Jamie laughed when he told me what happened next. He approached the calf to guide it back to its mother, and the tiny little Hercules head-butted him from sheer bravado, conscious, no doubt, of being the only male in the 'jungle', and fiercely protective. We are looking forward to watching his reaction when he finally meets all the other calves on the farm, all of whom are larger than he is.

ON BEING SKITTISH

No, not a new nationality, but a state of being when the wind has a particular and hard-to-define quality. Cows and sheep, horses and humans, hens, pigs and cats are all affected by the wind in very similar ways. Some winds can make you bad-tempered. Some are liberating and invigorating. And then there is Shelley's 'wild west wind', which leaves the cattle and sheep here certainly (because they never hide their true feelings) and the people here possibly (because people so often do) feeling skittish.

On one such day I took hay to the Beech Tree Field in a small trailer and instead of waiting for it to arrive and ambling or hurrying towards me, all the cattle and sheep cavorted, rushed and leaped in joyous enthusiasm prompted not just by hunger. The surge of youthful playfulness and the speed achieved saw animals overshooting the trailer, then, with a silent screeching of four-footed brakes as if they had hit a patch of black ice, wheeling round and returning to the hay in a whirl of leaps and half-somersaults.

CHRISTMAS

A lovely day of indulgence. With Gareth back in Wales spending Christmas with his mum, Rich and I have indulged in work and idleness in perfect proportion. I rounded off the day with a walk in the deep, dark wood, to find the two house cows.

I guessed that they would be lounging around behind one or other of the twenty thousand or so trees, waiting for me to find them. We wandered slowly homeward, they forging a true path through the black night, and me ducking and weaving like a good rugger-player, trying not to get my head knocked off by low branches, pretending or at least wishing I didn't need a torch. Then to the cow pen and their daily dose of bribery: hay and sweet apples in return for a gallon or more of milk each. They seemed happy and I know I was!

On the third day of Christmas I walked the two cows home for milking in near-total darkness again. Nell, uncharacteristically, dragged her feet (all four of them). She decided she couldn't walk another step without a few mouthfuls of horrid-looking old stinging nettles, which she consumed without blinking. Apparently blinking is good for one's eyes but I don't know if cows know this. Like most animals and certainly all wild ones, cows and sheep know instinctively when they need something extra in their diet. I wonder if, when humans crave certain foods, it indicates a genuine deficiency.

Gareth had returned from Wales early, and invited me to celebrate New Year's Eve with him. We worked the day together on the farm. Alongside his instant rapport with the cattle and sheep he is patience personified and he also, fortunately, has an unquenchable determination to mend everything, as well as the skills required to do so. I find at every turn that he can do jobs far better than I can. I didn't believe it possible that I could ever be so happy.

WHAT NO MAN LEARNED YET

The skies above the farm are thronged with vast flocks of jackdaws each evening, hurtling about enjoying life to the full, especially if the moon is out to play with. Joining them is a small elite display troupe of Red Arrow-like starlings cutting geometric patterns above the ash trees. Our small resident flock of starlings hardly ever visits the bird table, preferring to reside in the barn owls' oak and forage for food in the fields and hedgerows. The solitary pair we first saw has multiplied, and a quick, whooshing-past head count gives the impression of sixty, all of whom join hands to sky dance. Something to wait for and watch for and sigh for.

Birds know it all: if and when it is going to rain; where to find the best food; where and how to build their nests; when dire weather is on its way; and how to show off on a Saturday night. The daytime belongs to the buzzards and ravens, both angling their finger–feathers to manoeuvre like gliders, and to the ever-singing goldcrests, friendly robins, enchanting tree creepers, valley-dipping, prehistoric-looking green woodpeckers, smart, hatted willow tits, and all their friends. I recall the words of Edward Thomas in 'Sedge-Warblers': '. . . the small brown birds / Wisely reiterating endlessly / What no man learnt yet, in or out of school . . .'

As a sweet pledge of Spring the little lambs
Bleat in the varied weather round their dams . . .
While the old yoes bold wit paternal cares
Looses their fears and every danger dares
Who if the shepherds dog but turns his eye
And stops behind a moment passing bye
Will stamp draw back and then their threats repeat
Urging defiance wi their stamping feet
And stung wi cares hopes cannot recconsile
They stamp and follow till he leaps a stile . . .

– John Clare, from *The Shepherd's Calendar*

STAYING YOUTHFUL

Some people still go dancing in their eighties and nineties. Well, some sheep do too, in a manner of speaking. Spend every day in the office (graze the field); settle down, buy a house and bring up children (ditto, stay in the field and ditto); go back to work (keep grazing); play golf (enjoy an occasional skip round the field); retire and take up knitting (keep grazing but rest more frequently); or never dream of giving up work (graze enthusiastically); cram dozens of hobbies into a busy schedule (skip and run, jump and 'dance' every day); and, at eighty or ninety, go dancing with your grandchildren's generation (seek out the company of younger sheep and lambs and feel less than half your age).

Tealeaf always retained a keen interest in current affairs. At ten years old, she couldn't understand how her twin sister and their friend could be content to fritter away the days dozing and day-dreaming, so she upped sticks and left home. After one unsuccessful attempt at being admitted to the lamb gang, she finally became hip enough, and took to hanging out in the many and various dance halls of the high pastures, happily 'growing old disgracefully'. That's a phrase I will forever associate with Mary Cooper, co-author of a book of the same name. Mary was Mum's friend for seventy-three years from the time they met at school, and has been my friend and inspiration all my life.

MIDSUMMER MADNESS

On a midnight midsummer walk, Gareth and I were rewarded by an unexpected encounter with six emerald-jewelled glow-worms, scintillating in the soft dark grass. These predatory beetles have declined dramatically in numbers as a result of habitat loss, pesticide use and, I suspect, the annual release of fifty million pheasants, ten times more than in the 1960s. Their disappearance could also be contributing to the decline in nightingales, which feed on them. Yet, if they were encouraged, as they should be, they would play an important role in controlling slugs and snails. The molluscicide metaldehyde was widely used on arable farms. It is difficult to remove from water but was finally banned in 2022. On one occasion it was found near a drinking-water treatment plant on the Thames at forty-two times the official limit.[*]

In *The Duchess of Malfi*, the Jacobean playwright John Webster summed up the foolishness of human ambition (which today aims for big combine harvesters, big fields and few hedgerows, with even those that remain trimmed to oblivion) when he wrote:

[*] Castle et al. 2018, 'Measuring metaldehyde in surface water in the UK, using two monitoring approaches', *Environmental Science: Processes & Impacts*, Issue 8. Available at https://pubs.rsc.org/en/content/articlelanding/2018/em/c8em00180d.

Glories, like glow-worms, afar off shine bright,
But look'd to near, have neither heat nor light.

On 14 July 1667, Samuel Pepys took his wife for a day out in the country, by coach and four. As they were walking on the Epsom Downs, Pepys spoke to a shepherd and his son who looked after eighteen score sheep, earning them four shillings a week. 'He values his dog mightily, that would turn a sheep any way which he would have him, when he goes to fold them . . .' Later the same day: 'Anon it grew dark, and we had the pleasure to see several glow-worms, which was mighty pretty.'

The more Gareth and I walked, the more we could discern. The tawny owls hooted and a badger nearly tripped us over as we made our way across the soon-to-be-soaked fields to the slopes of the bank where the dropwort grows. By the light of a tiny torch, we gazed at the feathery white flower heads and tight, pink buds.

The next day, while bringing in a small group of cattle, I found, for the first time on this farm, an exquisitely delicate, coral pink grass vetchling.

BIRDWATCHING

Families of kestrels are having fun today, accomplished adults and youngsters with L-plates on, swooping from hedge to hedge and tilting their wings to the sun, with a Spitfire playing in the clouds above them, its Merlin engine sounding as sweet as the day it was made.

The buzzards were hogging the limelight during haymaking, placing themselves strategically while we mowed, plunging down into the swathes and returning uphill to a high tree with claws full of grass and a field vole, like sea eagles fighting their way out of the water with a struggling salmon in their talons. Two linnets were busy in an alder tree. Then the treat of a close encounter with a yellowhammer that flew slowly (can a bird fly slowly?) from a blackberry bush to sit on the fence, seemingly unaware of me as I meandered erratically around, pulling up ragwort by the roots. This is the one weed we don't allow to flourish, because of its poisonousness to cattle and sheep – especially dangerous in hay where they can't avoid it. A host of harebells stood out among the deep purple heads of dwarf thistles: delicate, translucent and determined.

TALKING TURKEY

One August morning I drove to Burford for a rendezvous in a car park with a breeder of organic turkeys. We could have been exchanging any number of illicit goods, but I received a large box containing twelve poults, handed over the agreed amount of money and drove happily home. As we installed them in their new quarters, Gareth whistled quietly, which seemed to calm and reassure them.

A word to the wise. If you find yourself weeding the cabbages, bent double under the expensive netting that is supposed to keep out absolutely every unwanted species, but of course doesn't, because of the one little corner you didn't pin down adequately, and if among the handfuls of overgrowth made lush by excessive rain you find slugs that you decide to capture and feed to the turkeys as part of their high-protein diet, don't forget to place a leaf or two in the box for the slugs to feed on. It will stop them all climbing back out while you are intent on pulling out a particularly tenacious dock.

Three weeks after their arrival, the fox took one of the turkeys. In desperation, we put the remaining squad in the old livestock transporter for safety, while I worked on a masterplan to allow their liberation in a better-fenced paradise. They had plenty of room, a perch, lots of food and water, nice bedding and bouquets of interesting weeds and grasses to nibble but after a day and a half, they made an unbelievable escape.

There was one (slightly) weak spot, high up above the door. They had obviously consulted, cased the joint thoroughly and maybe stood on each other's shoulders to push out the puny plastic window bars before flying to greener pastures. When I arrived, open-mouthed, aghast and lots of other synonyms, they were walking about in elegant synchronisation, pecking at this and that, and I was, of course, won over. Any further incarceration was clearly indefensible, so from then on I had to spend my days alternately gardening and fencing to deter Reynard's return, whistling to myself and my companions as I went. What better way to spend the next three months? At night, we shut them in a fox-proof enclosure just to be sure.

A month later, a fox climbed the nine-foot fence, squeezed between two lines of barbed wire, killed ten of the remaining eleven turkeys, and left the eleventh standing and staring in disbelief. The fox made a run for it as I arrived, and I saw him climb the fence, throw himself over the top like a commando and disappear. I picked the remaining turkey up. He was very heavy and I confess that I gave brief thought to the very expensive food they had all been consuming for so many weeks. I put him on the passenger seat of the Land Rover and took him on a trip round the farm, hoping to take his mind off the tragedy. The plan failed. Un ar ddeg (Welsh for eleven) was inconsolable, until he spied our four hens in the garden and took a great shine to one in particular, following her everywhere. She tried to ignore him, but that was never going to be realistic considering he was ten times her size. She was, though, definitely the one in charge. He watched her watching a spider but as he moved closer she pecked him hard and he retired obediently. The cats were more successful at ignoring him, but they all shared milk and other offerings, and sunbathed together.

NEVER TOO LATE

Having grown up on a remote Cotswold farm with a father and brother who could drive anything, it seemed perfectly natural that I should develop a healthy interest in farm machinery. The first piece of equipment I remember watching in action, sometime around 1959, was a yellow baler owned by a neighbour who contract-baled hay for us. This amazing machine produced small, solid, tube-like bales tied with wire, which were particularly user-unfriendly.

Over the years we have owned a number of balers: New Holland, Bamford, Alice Chalmers, International and John Deere. They all produced rectangular bales, preceded by the familiar churring as the pick-up swallowed the rowed-up swathes. The driver, looking behind more often than in front, was constantly ready to slam a left foot on the clutch to prevent the pick-up jamming with too much hay, and the consequent palaver of replacing the shear bolt if you weren't quick enough.

When the driver was me, I was ever mindful of the cartoon permanently fixed in the tractor cab, of a farmer baled into a bale, with his head and feet sticking out at each end. This image, coupled with my father's constant and absolute commitment to safe working, as well as his lectures on the dangers of the PTO (power take-off) and the ability of the pick-up to drag me inexorably in and through the shunting mass of metal spikes, slamming presses and knot-tying bills, gave me a heightened awareness I am grateful for to this day. It was perhaps Richard who afforded me the most learning opportunities, though. If he milked a difficult cow or drove an old tractor with dodgy brakes down a precipitous hill or a combine harvester on a banky field, I wanted to do that too.

My father and our next door neighbour went every year, with two tractors and trailers, to fetch straw from a farm a few miles away because none of our fields were suitable for growing corn. 'Never stack the same load twice' was a maxim drummed into me at an early age. There was a protocol or knack to building a load of hay or straw with small bales so that none of them would fall off, though on occasion a whole load could slide askew in an uneven gateway and would have to be secured with extra ropes or occasionally re-stacked completely.

As soon as Richard and I were strong enough to stack, loads could be completed much more quickly, with two strong men pitching bales up to us with two-tined forks until we were seven layers high. My first solo effort was when I was ten in 1963 and I created a hidey-hole in the middle of the top course in which my friend Madeleine and I could crouch, enjoying the illegal journey home, ducking as we passed under overhanging branches on the deserted side roads. Increased loading speed meant two trips in between twice-a-day milking instead of one. If you didn't have jeans with worn knees, you hadn't lived. I can imagine the reaction of my father's generation to designer-ripped jeans!

A few years ago, Richard decided to buy a JCB Loadall. He took Gareth and me to the local agent, but I was not impressed. The agent zoomed round the car park, telling me how simple it was to drive and that all you needed to do was keep your foot on the brake the whole time. One foot on the brake and one on the accelerator; it didn't sound like my kind of logic.

Nonetheless, the Loadall arrived at the farm, and Richard and Gareth mastered it with ease. I bided my time and read every word of the inch-thick handbook twice, memorising the twenty-four grease points that need attention after every ten hours of work. I bought my own grease gun and when no one was looking, I drove the monster, in first gear, to the middle of a large field for a practice. I grew to love it. By the time winter feeding started, I was performing tricks with three bales of silage at a time that the Health and Safety Executive would not like to hear about. I decide that it's never too late to own your first JCB.

THE HUNT FOR RED NELL

My belief that cows deserve and need freedom is occasionally called into question. Gareth and I once spent an entire day searching for Red Nell, who was not the kind of cow to go missing without a good reason. We had to find her. It was cold, threatening rain, and she could so easily have come home that we felt certain she was calving and in need of help. The mobile phone signal in the wood is intermittent to non-existent; we went off in different directions and met up from time to time by chance. Eventually we agreed on a quick dash back to the house to fortify ourselves for the next shift as it was now raining steadily and getting darker by the second. But before we turned back we walked together up to the top wall of the wood, peering under every tree.

Gareth pointed at a small clump of densely planted, floor-sweeping evergreens and I smugly claimed that Nell would never dream of hiding in such a place. He looked anyway and there she was, curled up like a cat, with resignation written on every strained muscle of her face. We were so high up by then that we did have a phone signal and could ring Rich with precise instructions. Only he could have managed to get a vehicle to such an outlandish spot. He arrived with calving gear but there was no visible sign of a calf and we made a joint decision to try and get her home.

We appealed to her extremely good nature, asking her to muster the strength to walk back down the steep, rutted tracks to the barn. Rich drove behind to supply light and Gareth and I walked on each side of her, making encouraging sounds. She was okay, and to my great surprise walked remarkably easily. Once she was home and we could see her clearly, she seemed more bad-tempered than in pain, though one does cause the other. Richard rolled up his sleeves and investigated. He found the calf's feet, pulled gently and a healthy bull calf appeared without much ado. Nell set about licking him dry with such angry energy it looked as if she intended to kill (or eat) him. She suddenly noticed another calf, who was already in the same pen with her mother, and set about licking him ferociously too. We offered her water but she was still angry. In our exhaustion, it took us several minutes to realise that she was going to have twins. She marched round the pen twice, lay down and calf number two, a heifer, slid out with ease. The two calves must have been a bit tangled up to start with, but sorted themselves out as we walked her home. Nell relaxed and licked them both gently. She drank three full buckets of water and was her old, genial self again.

To every cow belongs her calf, therefore to every book belongs its copy.

– King Diarmait Mac Cerbhaill, *c.* AD 560

A SHEEP LEARNING CURVE

Our lives were for many years dominated by cattle, but in the 2010s we began to build up our embryonic flock of sheep to semi-commercial numbers. The first year of lambing a big flock was an exciting, worrying and immensely tiring time. We gave it our all, going to bed late, checking them in the middle of the night and getting up early again. In fact, we probably gave it too much; most of the ewes didn't need us at all, but we would have hated to fail the few that did, so we prowled and peeped and wore ourselves out.

The next year, we were still on a sheep learning curve but we felt much better prepared. Our first lambing season had taught us a huge amount, and we had three expert neighbours who were generous with their advice and offers to help 'at any time'. Sheep are far cleverer than we are so we gave them space and time and merely loitered in case we could be of use. The first time round, all the ewes had just been numbers but now we knew them as individuals, and there were little signs that alerted us to those who might need help. Aesop knew how easily a young lad might get bored watching sheep (and crying wolf when no wolf was there) but a true vocational shepherd, happy to live with his sheep, would understand them and perform wonders in helping them overcome obstacles and dangers. The weather was so unpredictable that we constructed instant lamb villages out of abandoned Land Rover canopies. Lambs are very quick to learn and will happily accept offers of help or shelter if they feel they need them.

I was a bit surprised to discover a dangerous sheep last year. They appear to be soft woolly things who don't bite or kick and who could stand on your foot without you noticing. One ewe, however, seemed to come to the conclusion that people were to blame for the pain she felt giving birth to twins and she tried hard to demolish two of us with head-butts.

In his book *Farmer's Glory*, A. G. Street tells us how the shepherd was by far the most important person on the farm and the whole farming year revolved around the sheep. One year when Street's father mentioned that he had agreed to sell a rick of surplus hay, the shepherd told him he would leave if he did. The sale was cancelled and the shepherd remained.

MILK AND COMFORT

I imagine most farmers know that cows and sheep groom their offspring with their teeth as well as their tongues. This was filed away at the back of my mind until two events reminded me of it.

One of our cows gave birth to twins, but after licking them both dry she chose to keep one and utterly rejected the other. We've had a suckler herd for nearly five decades, and something like this has happened only four times before. We tried every trick we knew to persuade her to rear both calves but her obstinacy was greater than our cunning. I had to spend a considerable time each day grooming between one and five other freshly calved cows with brush and curry comb in order to distract them while the rejected calf was able to suckle one quarter from each cow. As a result, he got ample milk, but none of them groomed him; that was my job. In return he groomed me, with his tongue and his teeth, carefully pulling at my trousers (and the skin beneath). He also licked my hands and arms with his rough tongue, but I kept my face out of reach.

On the ovine front, one of the five ewes to give birth to triplets that spring produced a lamb that was tiny, wool-less and to all appearances premature (though his two siblings appeared full term). He was rapidly becoming cold and stiff as the ewe was licking the bigger lambs first, so I rushed him to the house, wrapped him in a woollen cardigan and placed him on a hot-water bottle by a radiator. He gradually came to life as he warmed up, and I gave him very small quantities of colostrum from a syringe.

Providing milk and comfort was painstaking but possible. Grooming such a tight-skinned titch was not. By day three I felt I could risk taking him back to his mother, joey-like in my coat, in the admittedly rather forlorn hope that she might accept him. She wouldn't even look at him, and nor would any of the others, except one. This seemingly compassionate ewe saw his weeny head peeping out of my coat and gave him a kindly lick, before going on to bite and tease the stiff skin on his head and neck with incredible delicacy into a soft and comfortable state. As she did this, I could feel his whole body relax with pleasure. I placed him on the ground and she continued grooming him until every bit was done. In the space of a very few minutes he was transformed from a crumpled object into a lamb with self-esteem.

One of the benefits of still having a relatively small flock of sheep is being able to know them as individuals. After one lambing season we merged the ewe lambs born the previous spring with the recently lambed ewes. Partly we did this to increase the proportion of adult sheep to lambs, which reduces the need for wormers – older sheep can acquire natural immunity, baby lambs have little or none. The other reason was to shut an additional field up for haymaking, even though we knew it was late in the season and we could expect only a small crop. The lambs had not seen their mothers for the past six months, half their lifetime in fact, and it was great fun to see them become reacquainted. The interaction between the grown-up ewe lambs and their little siblings, however, was varied. Some showed a definite interest in them, some just gave them a fleeting glance and carried on grazing.

We talk about people behaving like sheep, which assumes that sheep all behave in the same way. That has not been my experience. When frightened, of course, that's what they do. But in general, some are affectionate, others prone to head-butting. Some are determinedly self-sufficient, others seek our help when they need it. And some can be trusted to lead the flock home. They are as individual as we are.

The reintroduction of the female lambs to their mothers provided a good example of this. One of the ewes was a loner. When I went back to check on them a few hours after mixing the two flocks I couldn't initially see her. I eventually found that she'd taken her two current lambs and also her newly reappeared yearling daughter through the open gate into the next field. They were all sitting together under a thorn bush at the far side of that field, almost half a mile from the rest of the flock.

AN AUSTRO-GERMAN KNITTING CIRCLE

For a long time the entry for 'holiday' in the (theoretical) *Kite's Nest Dictionary* said 'definition not available' but all that changed with Rich's trip a few years back to a part of northern Italy not far from towns featured in so many of Shakespeare's plays. He was there for his health, at a Rudolf Steiner-inspired anthroposophical retreat, walking, swimming and consuming fish, salads and fresh vegetable juices. He sent an email that read:

> I have joined an Austro-German Knitting Circle – a group of friends I've made, several of them from Steiner establishments in Germany. They are all women between the ages of about fifty-eight and seventy who knit. The Austro- in this case is Australian not Austrian! One of them is leaving tomorrow and she wanted me to go to a little leaving party she was throwing tonight. To be admitted, you had to take some knitting that you'd done. I took my Kite's Nest woolly socks, to their great amusement, and claimed I'd reared the sheep and shorn the fleece, although my mother had knitted the socks. That got me in nicely. One of them said, 'They still smell of sheep.' I said, 'That's not sheep; it's probably my feet', and they all dissolved into much laughter!

ALLERGIC REACTIONS

We once had a very enthusiastic young agricultural student, Ben, on placement at the farm for a few weeks to help with the lambing. He had a severe nut allergy, which caused no problems but led me to wonder whether any wild animals ever have allergic reactions to food. I would be inclined to think not, as long as they are allowed the freedom to roam and choose what they eat; animals confined in zoos and pets in houses might be a different matter. It seems increasingly likely that humans have caused the problems that cause humans to suffer. Intolerance of gluten has escalated in tandem with the deliberate breeding of wheat with higher amounts of gluten. Wheat varieties have perhaps developed more quickly than people's ability to adjust to the change.

The public buys its opinions as it buys its meat, or takes in its milk, on the principle that it is cheaper to do this than to keep a cow. So it is, but the milk is more likely to be watered.

– Samuel Butler, from *Notebooks* (1912)

JUMPING FOR JOY

Every now and then a sight on the farm feels like a real stroke of luck. One such event was the tiny new-born calf I once saw jumping for joy like a lamb. Nothing unusual, you might think, but I knew there was a far greater significance to my observation than the calf merely being happy. It meant that, without any doubt, the calf had suckled milk from its mother, its very first and most vital drink. Late the previous night, when the calf had been born, the strange 'talking' of its first-time mother had drifted down to the house from the windy hilltop and drawn us up to find the pair. The mother seemed unsure how to treat her calf and every time he tried to grasp a teat, the cow, seemingly wishing to keep him in her sights, turned full circle. She talked to him constantly in a strange, disturbed monologue and this drew the other members of the herd over to investigate, making it even harder for cow and calf to get to know each other. We managed, by a stroke of luck, to open a gate and let them slip through into a field of their own. We were concerned that the calf had not suckled, but decided to leave them be.

Later on, I was walking slowly and quietly, stopping often to look at the sheep. I saw something moving the grass in front of me and froze. Two short-tailed field voles emerged, nibbled a buttercup leaf and then walked across my boots and played on my laces for maybe ten seconds. I did not breathe.

WHITE DOT

One crisp autumn day, I walked out of the back door and up the path and was surprised to see White Dot coming down the road towards me at speed. There were no other cows in sight. I rushed to meet her and opened the gate by the side of the cattle grid. She hurried past me. The gate to the collecting yard happened to be open and she marched in and installed herself in the crush. I already knew there must be an urgent reason for this behaviour. I had cultivated a friendship with Dot when she was a young heifer, feeling that I might one day need to be of use to her; as her mother was the only Dot ever to be defiantly unfriendly, I knew that her daughter might well follow the same pattern. That day she looked and felt unwell. She had a swollen quarter infected with mastitis and she needed antibiotics, painkillers, bathing and hours of attention.

Hannah Steenbergen, a colleague of Rich's at the Sustainable Food Trust, was staying with us at the time, as they were working on a project together. She not only offered to help on the farm, she actually did help. Over the years, so many people from every possible background have offered to help but very few have been truly able to be useful. Of course, that is not surprising. It is extremely hard to define what it takes but Hannah, fortunately, possesses 'it'.

She is shorter and slighter than me but under Rich's tutelage, she quickly learned how to drive an enormous tractor and control its potentially dangerous heavy trailer on a steep hill, with aplomb and competence. It took me a lifetime.

White Dot calved, slightly prematurely, and for the first few days her tiny white daughter was satisfied with her mother's meagre store of milk in her three unaffected quarters. On about day four, Hannah sensed the calf needed more, and came to tell us. Rich taught her how to warm the milk to the correct temperature for the calf and we were both stunned by Hannah's absolute confidence, implied but not expressed, that she could teach the calf to suckle from a bottle, a not-usually-simple task. I felt I simply could not voice my doubts. She succeeded magnificently and as if she had been doing such things forever.

Later that evening I found myself on a forty-three-year-old tractor in the middle of nowhere (or maybe somewhere), surrounded by quickly darkening skies, darkening cows and ever-luminous sheep.

DANDELION

A very small lamb called Dandelion was born on 16 April 2018 and my world has been enhanced by her ever since. Gareth and I were in the bottom paddock when one of the Hebridean ewes gave birth to twins. She licked them both dry but when they went to suckle, one each side, she pushed Dandelion away while allowing her twin brother to drink. Dandelion tried again and again, but the ewe pushed her more aggressively every time. We knew we had to intervene.

Dandelion weighed one and a quarter kilos and her tiny mouth was too small for a normal lamb teat so I gave her colostrum with a syringe. She was quick to learn and sucked enthusiastically on the hard plastic as I depressed the plunger as steadily as possible. Two days later I found a minuscule bottle and teat online; they pleased her so much she went round and round in happy circles. She was very small and very determined, lion-like.

She needed small and frequent feeds, so we took her with us when we went round the farm in the Land Rover, with her 'picnic basket': flask of hot water, cold water too, just in case, milk powder, jug, whisk, bottle and teat. She was lifted in and out, and this infinitesimally small lamb was completely unfazed when surrounded by enormous sheep or gigantic cattle, standing her ground while they mentally weighed her up. Often she followed us, watching whatever we were doing. When she got tired she slept on the middle seat, on my favourite coat. She helped us move the cattle: they followed her out of curiosity as she followed us. She helped us move the sheep, as a trainee 'sheepdog' by our heels. She lived in the garden and often slept on the doormat.

To begin with, the cats were not sure they liked her. When a lamb greets another lamb, it does so with a gentle head-butt; Dandelion head-butted the cats and the cats swiped at her. They very quickly came to an understanding and agreed to share the mat. In fact, on the day Dandelion decided she no longer wanted to run to come with us in the Land Rover, two days before she was ten weeks old, she stealthily walked in the opposite direction side by side with a feline accomplice, looking over her shoulder to see if I was watching. They spent the day together, Dandelion nibbling grass as the cat sunbathed.

She was and is a lovely friend but, fortunately, unlike some pets who imprint on their human saviours, as soon as she discovered grass she no longer needed us, and although she runs to greet us still, she makes it clear that she can't spare more than one minute from her full-time grazing obsession. And, unlike all previous hand-reared lambs, despite me taking her to 'play school' so she could interact with other lambs, she is so totally self-sufficient that she actually appears not to like or need the company of any creature at all. She knows how to ask for a gate to be opened and for hay when summer grass begins to lose its sweetness and sometimes she will race the Land Rover home for the fun of it, but often she will ignore it, and me.

She's still here, currently doing whatever she fancies with her daughter, Carnation. They lie on the road outside our house, 'calming' the traffic better than any sleeping policeman. I have spoiled her. She approaches most of our visitors but usually ignores me, though when I recently broke a rib after a fall from a height and was standing on the farm road, unable to walk for half an hour, she showed definite concern. She marched over to me and touched my hand with her nose, seemingly asking if I was ok. That's all the conversation I need, actually.

Thinking about how lambs greet each other with their heads, there is really not much else they could do. This is why I never treat a ram or a bull as a pet in the way you can with a dog. If you stroke a ram or a bull on its head it almost always enjoys it but, of course, it will want to reciprocate. Even if it merely wants to ask for more of the same, a ram or a bull can reasonably use only its head on us, as it would on one of its peers. The trouble is, they very quickly become far more powerful than we are and what they may intend as affection appears and certainly feels like aggression. We would then have to avoid them, which they couldn't possibly be expected to understand.

Over the years I have been unwittingly carrying out experiments into sheep behaviour; I expect all shepherds are doing the same and our joint observations would provide the source material for a full university course. Dandelion and her daughter have appeared supremely happy for the last year, but recently a group of lambs the same age as Carnation came down to the fence one day and seemingly persuaded her to 'leave home'. She leaped the fence and danced off up the hill surrounded by her new companions. We are all busy trying to work out what Dandelion is thinking. It's so easy to anthropomorphise, and I do so often, but only to describe the things I see to other humans in a language they understand. She looks a bit lost, lonely, bemused, offended even, but she hasn't followed her daughter as she could have. She doesn't like sheep except her own lambs. I often wonder if she knows she is a sheep, after I reared her and she spent the first ten weeks of her life consorting with our cats. People I know say they 'like their own company'. I think she does too. After four days on her own, patrolling our farmyard and the paddocks to which she has access, she chose to go with the very newest lambs and their mothers to the Lake Field, far away from the direction Carnation had gone. Our observations and ruminations continue.

Mary had a little lamb
  Whose fleece was white as snow
And everywhere that Mary went
  The lamb was sure to go . . .

Why does the lamb love Mary so?
  The eager children cry;
Why, Mary loves the lamb, you know,
  The teacher did reply.

– Sara Josepha Hale, from
Poems for Our Children

MAKING HAY

At the height of summer I am normally to be found hurtling round the field on the Ford 4600 with a small but handy haybob, tossing the neatly mown swathes up in the air to land all over the place. It's an age-old process with many names – tedding, spreading, throwing out, turning. The scattered grass will bake and then the following day, as soon as the dew has disappeared, I will ted it again to bake on the other side like a biscuit (bis cuit – 'twice cooked'). Once it is crisp and I can pick up a handful, twist it tightly and feel it crackle, I will change the direction of the tines to make the machine row it up ready to be baled.

Haymaking has always been the big event of the summer. I hated being imprisoned in school when I knew the grass was being mown.

Oddly, the excitement of haymaking time never dims. I say 'oddly' because it is a time of unremitting hard work and often a great amount of worry. But making hay or haylage or any of the other ways of conserving grass for winter feed is an integral and essential activity for us. More often than not rain threatens soon after the hay is baled and we work late into the night to load and store it safely in the barn.

For the past few years, though, there has been no rain at all. The sun takes centre stage every day and the heavens are as blue as they were when I first started at secondary school and everyone lay on the lawns and playing fields during breaks between lessons and stared at the sky.

After a day of haymaking I return from looking at the cattle via the wood and decide to park the Land Rover at the edge of the woodland road to watch the proverbial world – in this instance, the insect world – go by. Bright butterflies up-tumble to the tops of the tallest pine trees in ones and twos, and Lewis Hamilton-imitator insects drive their invisible racing cars up and down the woodland rides. They all have such purpose: an aerial insect-way with no roundabouts or Highway Code, and no accidents.

DAPHNE

I emerge from the wood into a sunlit field to see minuscule lambs climbing on the uneven gathers of the old oak's trouser turn-ups, and marvel at the eternal springs in the tiny legs of lambs.

Much later, Dandelion and I do the rounds of the Hebridean flock. They have all settled down for the night and allow us to weave in and out and sit among them as if I too were a sheep. Dandelion has sufficient ovine credentials, even though she left home at a few minutes old.

Chocolate Brownie, a tiny lamb just five days old, fights a plastic fence post, which bends and then springs back upright. She reverses repeatedly and head-butts it, then, slightly off target, she butts the electric wire and feels the shock. She seems to assume it must be the fence post fighting back, so she renews her assault with gusto.

In the heat of July, ewes and lambs swarm across the recently mown field to seek the shade thrown by the big bales of hay, though their shadows are pitifully small. The sun roasts them through the hot blue roof and when the bales' shadows diminish as the sun makes its undisturbed trek from east to west, the sheep creep round to the other side.

A white sheep called Daphne gives birth to a tiny ewe lamb (yes, I know it's July). She is a silvery blue with an almost black head and enchanting cream make-up near her eyes. Little Daphne is extremely hot and Daphne licks her to keep her cool. She licks her face and head, and she licks her legs. Four days later, as we are bucketing out a water trough, I notice Little Daphne try to sneak some milk from another ewe. The ewe walks away from her. The lamb then comes to 'speak' to us – loudly. It isn't a bleat or a baa or a cry; it is a lamb noise, signifying a question. We are very hot and very busy and although we feel pleased to be spoken to, we merely reply with 'Hello'. She turns and walks to the Land Rover, which is parked a few yards away, and speaks to it in the same questioning manner, though more loudly. We realise that she had been asking us for something and, seeing we were too dim to understand, had gone on to try asking the only other non-sheep object she could see.

She must be hungry. And she must be very hungry to be bold enough to come to ask. I pick her up and feel her tummy. She feels quite empty. I place her back on the ground but hold her so she can't walk away and, as I had hoped, Daphne walks over to see why I am restraining her lamb. I grab hold of one of Daphne's horns, letting go of the lamb at the same moment and then cuddle Daphne, releasing her horns. Our sheep permit brief, horn-as-handle tricks but actually hate it. Daphne's udder is small and inadequate.

We load ewe and lamb into the Land Rover and drive home. They spend the next few weeks in the walled garden of the house and sometimes in the orchard. I am bottle-feeding Dandelion, so I give Little Daphne supplementary feeds as well. She still suckles from her mother but I can't help feeling that Daphne is becoming even lazier about producing milk now she can see that I am prepared to help.

Some sheep just decide to be friendly; sometimes this can be a very big surprise. Old Cocoa was 'just a lamb' until one day, aged around seven months, she came over to be stroked. From then on, she was always approachable. Every time we rounded up her flock, she would choose to walk by our sides, collie-like, till the last sheep had entered the field of destination, before trotting through herself. And then her daughter did the same and later her granddaughter. So now it is not unusual to have three collie-like sheep helping to round up the other sheep.

Hand-reared orphan lambs have a different outlook on life and almost always remain more friendly to humans than ewe-reared lambs. The occasions when an 'ordinary' sheep (no such thing) walks over and starts a 'conversation' are more moving and valued.

TREES BE COMPANY

With the arrival of autumn, yellow poplar tops pierce the sky. Hens and little Hebridean sheep colour-match each confetti leaf of walnut, beech, oak, ash, hawthorn and lime as they are all blown this way and that.

A few years ago I saw an article on the BBC website giving nine ways to feel less lonely, but it did not include the words 'go outside'. That would be the first step I would recommend.

Nature in all its forms, but perhaps most of all trees, can provide an actual physical presence that one can relate to. As William Barnes wrote:

> *However lwonesome we mid be,*
> *The trees would still be company.*

One crisp October day, the leaves of the mulberry are rustling as I climb up into the branches to pick the more-than-generous fruit it still offers. The sweet wind crackles the leaves and I know that the tiny spots of rain we had did not convince the fruit trees and bushes not to worry. The trees know what to do when there is a drought: close down all systems and slip into autumn mode, producing monumental quantities of fruit to ensure the next generations in case they should perish. For the first time in many years, although the squirrels are working their socks off, gathering and storing walnuts and hazelnuts, there are so many that I can gather some without feeling I am stealing from them, and that's a treat.

Apple, pear, hawthorn and walnut trees are laden with fruit, and I avoid walking directly beneath them for fear the branches will give way. Gathering and carefully storing what's available now provides a bulwark against the need to buy fruit from afar during winter.

Most things in my life are solvable with hard work and the answer to almost every problem is to work even harder, physically that is. But any problem not in this category can be addressed by climbing a tree.

This particular mulberry was planted by my brother Richard some four decades ago, as a direct result of reading that King James I had required landowners to purchase and plant mulberry trees (at the rate of six shillings per thousand) to fuel what he vainly hoped would be an English silk industry. Richard fenced the tree so well that no cow, sheep, deer or rabbit has ever hurt it and the same fence is still rock solid today.

I climb ever higher, knowing that if I stretch for one of two berries in close reach, the second one is sure to plummet to earth. I reason that, should the branch I am standing on break, the chances are it will drop only as far as the branch below it and jolt me, and that I won't actually die for my exhilaration.

I go for one last blackberry hunt before the frost and realise it would take too long to find enough for a single pie, so I leave them all for the birds. I think about the dormice who ought to consider this particular hedge ideal, festooned as it is with hazel, honeysuckle and a continuous hand-holding parade of hedgerow trees: ash, oak, maple, plum, hazel, willow, hawthorn and sloe.

I take an unaccustomed shortcut to get to the Rickety Rackety gate and surprise a green woodpecker, which startles away yaffling. Two strides later I disturb a buzzard, happily and (as he thought) safely occupying a tree. He glides away unabashed and waits in the sky for me to pass, hoping to land again. Seeing me dither, he drifts sideways behind the branches and disappears.

The wild cherry trees are slipping into autumn colours. The willows near the stream are still a vibrant green but the oaks have darkened already. Sycamores are the colour of overripe broccoli and brambles appear to be growing a yard a day. The horse chestnuts are rusting quietly. Walnuts are a hard-to-describe green-grey and the elders look insolently youthful. Maples still wear shocking yellow scarves – bringing to mind the William Barnes poem 'Meäple leaves be Yollow' – and the hawthorn is laden with bright red fruit. Tall poplars throw down leafy twigs while trying to maintain an air of permanent verdure. Sheep reflect the sunlight; the black bull absorbs it.

Later, in the glorious early-evening light about half an hour before sunset, I walk the sunny but crisply cold fields of sheep. The sun's rays magnify the cobwebs covering the grass. I have only ever seen them like this in early-morning sun before, when I have marvelled at their sheer extent and worried that the constant grazing motion of sheep and cattle would destroy the spiders' brilliant constructions. I had assumed they must have to wait till quadrupeds slept to repair or renew their meal-nets. I now wonder if that process occurs constantly during the day, with patient spiders never allowing themselves to be deterred by the perambulations of heavy feet. I remember reading that an orb spider has to manufacture around twenty metres of silk to make a web.

This musing leads me to reflect that the activities of people, forever rendering some tiny species homeless, are an area of concern not sufficiently discussed. It is the tiniest creatures that do the most work and that work is totally vital for the continuation of life on Earth.

GAMBOLLING BY NIGHT

At fifteen minutes past midnight one late October evening I saw to my astonishment two lambs playing in the moonlight. The big moon had come up and seemed to linger for a minute on the horizon before rising slowly and lighting up the night sky. The lambs evidently felt frisky.

We'd finished supper just before midnight but then Gareth heard a lamb calling. I decided to go and investigate. At the very least, I thought, one of them must have got stuck in a fence, or was trying to wriggle under one. I homed in on the sound, torch in hand, in case the moon's rays were insufficient. Cry. Cry. I spoke to them: Dandelion? Daphne? And zoom, two missiles were at my feet to be cuddled. Nothing wrong at all; very unusual. I stayed for a little while, enjoying the affection; as I turned to walk away, they stayed by the orchard fence. Suddenly I was 'attacked' from behind. Daphne had run full tilt into the back of my leg, nearly throwing me off balance. I remonstrated and walked forward. They both followed in glorious, jumpy bursts of speed all the way to the gate. I climbed over and they began a game, racing and chasing each other in circles, ellipses and all sorts of other shapes.

What could I conclude? They had been as good as gold all day, eating, and now they wanted some fun. But why did they call again and again and again, until I went to see them? Well, I could invent a reason but, luckily, I shall never know for sure. They played and I saw them; that's enough reason and a perfect way to end my day. I am wondering how long they will continue.

A bleating sheep loses a bite.

– proverb, late sixteenth century

ALL IS DARK AGAIN

As November becomes December, it starts to get dark so early that we have to visit the sheep by torchlight. Their eyes are sparkling fairy lights of gold and nearly red, of icy blue and twinkly green, and of the silver that makes the stars. They blink, or we turn away, and all is dark again.

The last stretch of our night walk up the drive is stripes and bars as the moon throws the poplars' shadows across our path; we are illuminated and disappear every few steps. I think about the poetic beauty of Peter Pan asking Wendy to sew his shadow back on.

Christmas arrives, bringing with it a light covering of snow and a deliciously quiet day. The cows and sheep are fed, bedding straw is spread in seven journeys, two bales for each of four separate groups of cattle, and four bales for the fifth. In at 8.30 p.m. and a lie on the hard kitchen floor to recover before cooking a Christmas meal for Rich and me. Gareth is in Wales.

After supper I walk round the sheep carrying secateurs, liberate six from the still leaf-laden blackberry bushes and then, inspired by *The Butterfly Isles* by Patrick Barkham, search several groups of blackthorn with torch and magnifying glass for eggs of the brown hairstreak butterfly. In vain. I spy two delicate crab spiders hanging from a single silken thread and one motionless rosy rustic moth.

The following day, the sheep are racing round the field. The snows of yesterday have been warmed by the sunshine of today and the fields are snow-striped. The sheep are not running to keep warm; they are having fun. The grass is fighting back and there is nothing the sheep are more pleased to see.

Grass is the answer to everything. The sheep know this instinctively, whereas we have to learn it. In the soil is life. Grass covers the soil to protect it. Together the soil and the grass shelter a wonderful web of interdependence where everything fits in its place. One wrong step and everything is thrown out of balance.

The south-facing side of every anthill – and there are thousands in the Humpy Dumpy Field – is warm, and the snowy north half is still cold.

Anthills remind me of the extraordinary life cycle of the large blue butterfly. Though since reintroduced, it became extinct in England because the precise and delicate balance between grazing animals, anthills and wild thyme was disrupted. All species rely on a whole raft of factors to sustain them but only humans have stepped out of line and tried to dominate and destroy, rather than valuing and co-operating. Various factors had precipitated the butterfly's decline and the last meadow stronghold was fenced to protect it, inadvertently excluding the grazing animals on which its life cycle depended.

What humans decide to do, as opposed to what animals do intuitively, affects everything. If humans decide to kill, cull, reduce, prune or remove one species, it will be at the expense of another, or several; a chain effect is set in motion with unforeseeable consequences.

SPRAWLING BEAUTY

Slow walking is a must for the dedicated wildlife observer. The only friend I know who possesses this particular skill is the artist Matt Greenhalf, who 'sees things' as he walks and files them away for later. One spring day I received a text message from Matt saying, 'I am at the big old fallen oak where I have found the most beautiful and large basking lizard.'

'I am on my way, don't move.'

I ran the first part then walked as softly as I could before tiptoeing towards the tree at an oblique angle. This ancient oak toppled over in 2011. I saw the magic sight of the sprawling beauty on one of the few branches still visible above the blackberry that was engulfing it. Matt knew I was beside him but didn't move or breathe. The lizard disappeared so suddenly we didn't see it go. We stared at the empty space. Matt moved his eyes, almost imperceptibly, to point at a spot just below the basking platform where, in a hanging fold of loosened bark, I could just make out one hand and one eye of a fully grown toad.

One oak tree, living or dead, provides a home for hundreds of species of insects and caterpillars, thus supporting birds too, not to mention beetles and fungi and the many creatures that eat the acorns. It also exerts an indefinable effect on people. I glanced towards the Walnut Tree Field and saw two big calves babysitting a tiny new-born calf. The two were not, in fact, big by any standards other than their own; they had been born three weeks prior. None of the mothers was anywhere to be seen.

THE PSYCHOLOGY OF FARMING

The other day I spent quite some time trying to persuade two sheep to go through a narrow gap I'd created by partly opening a gate. I succeeded only when I decided to give up and went to close it, which made them both decide to seize the chance to dash through.

Psychology plays a big part in farming, whether the farmer is aware of it or not. My young life was lived in a small, isolated village and I got to know many very old-fashioned farmers. They all had their own ways of dealing with animals: superstitions, beliefs born of experience, knowing when not to make eye contact was a good idea and when doing so was vital. But I doubt if any of them had even heard the word 'psychology', let alone been aware they were practising it.

Farmers who worked with horses would tell you exactly where to stand, how to approach each horse, exactly which horses could be trusted to lead, and which would follow. They could list all the things that would frighten one horse and not bother another. And the people who milked cows, twice a day, day in, day out, through illness and depression and happiness, they all knew how to move, how to speak, what to do and what not to do to get the best results and cause the least upset. In many instances you simply had to do things in the right order; anything out of sequence could result in a drop in milk yield. All this knowledge was, in fact, an absolute acknowledgement of the individuality of all the animals they tended. In those days every farmer I knew instinctively treated all their charges according to their unique natures.

The arrival of spring at Kite's Nest sees a hive of activity among creatures big and small: birds, bats, bees, fox cubs, not to mention lambs, who get together in packs to 'Derby' it round the field just before dusk. They would prefer to 'Grand National' it, if only we would construct a series of objects they could race up and jump off. We have lambs in colour combinations to leave Yves Saint Laurent speechless, and the early purple orchids hide behind and between the bluebells, which are themselves crowded by verdant wild garlic and dotted with red campion.

I read *Extraordinary Insects* by Anne Sverdrup-Thygeson recently and have been enjoying the farm even more as a consequence. Filling canisters from the spring-fed pipe is no longer just a chunk of time to be lived through, but an educational trip to a whirring insect world on, in and under the shallow water of the trough, with its ever-depositing silt, stones to hide beneath, cracks, crannies and niches. Insects everywhere – floating, drifting, swimming, diving, flying, crawling, wriggling, buzzing, humming, surveying and dreaming all round me. Pulling myself back to the present, I find I am almost nose to nose with twin Hebridean lambs, fascinated by my inaction.

AND THEY'RE OFF

A starting line has formed for a round-the-field race of committed seriousness. An acceleration of lambs running at break-neck downhill speed and no clear leaders. None deliberately seeking another's slipstream, all quite bunched to begin with. But what's this? Three of the biggest, fastest lambs do an unannounced handbrake turn (you can almost smell the burning rubber) and race back uphill.

The 'field' is left stunned. Should they cheat and try to catch up by turning now, or should they complete the course and win or lose with honour?

The ewes trot over for prize-giving but are left as confused as anyone. Then suddenly a London Marathon camaraderie takes hold and the entire flock takes off for a circuit: sprinters, galumphers, flyers, very fast walkers, lots of identical twins confusing the placings, ancient matrons hurrying with no idea why, some finding they still have a spring in their old legs. A clear winner ... who then decides a second lap is in order. And so they play. A few drift away. Some flop down and sleep, some bleat plaintively while searching for the right udder. Night falls slowly, the sun sets upside-down, appearing to fall from a cloud rather than sink into the horizon. And silence.

The hand-reared orphans, sweetly destroying the flower

garden, miss sports day but invent games of their own, changing their best friend hourly: Sookey, Wookey, Tookey, the triplets whose mother died, and Theophilius, twenty days older than they are.

Wookey, the boy, makes friends with Theo. They play. By the next day Wookey has learned how to jump but Theo is too heavy.

So now Wookey and Sookey play. They jump out of their field and sit one on each side of the front door like lions couchant. They then return and jump back in. Tiny Tookey is still too small to jump but she does hold the world record for speed-drinking from a bottle.

Tookey drinks, grows, learns to jump, and all the triplets set up camp on the doormat, putting the cats' noses very slightly out of joint.

Theo asks. We open the gate. Four lambs take over guarding our house and I don't have so far to walk to feed them.

Every time we gather the sheep to inspect their feet and teeth, greet them as old or newer friends and generally give them the once over, we make notes: ewe 267 old, tiny, brave, determined; yearling 685 pretty with three good feet (!); ewe 390 supersonic; ewe 129 rides shotgun, spotting the smallest opportunity to dash off and lead the whole flock away; lamb as yet untagged, daughter of one of the leaders, just like her mother; the Blue Ram smart and spry and really quite kind with excellent feet; Gentle Jane patient, friendly, ancient, totally lovely; ewe 191 crazy, foot-stamper, head-butter, dangerous, admirable, one-off, keep.

Sometimes out in the fields a newly born lamb will stay asleep as its mother grazes farther and farther and it might wake up alone. In the past, if I was close and could hear its lost cry, I would try to work out the identity and location of the mother. If I got it right, the lamb was all too happy to be carried or guided towards her, but if I chose the wrong mother, the tiny imp would absolutely refuse to be taken in the 'wrong' direction and would return, time after time, to where she had been. I have now learned to leave it to the ewe to come back when she is ready. After all, I know that deer return to where they have told their fawns to wait for them.

Learning lessons is obviously instinctive and vital for all creatures, yet it seems to me that it is only humans who ask to be allowed to make their own mistakes. Every animal I have ever observed learns life's most important lessons from its mother. This is of course yet another reason, if one were needed, for outlawing the forced weaning of very young animals.

Being only human myself, I admit that it took me a very long time and the enduring of some very long and fruitless searches before I realised that one never sees a new-born calf on the second day of its life, unless it lives in so small a field it simply cannot hide. Experience and exhaustion have taught me that a calf's first attempt to suckle from its mother is only marginally successful and that the small amount it consumes is just sufficient to give it the strength to try again after a short sleep. The second or third attempt usually results in bliss and hence over-indulgence. Thus day two is invariably spent sleeping it off. You will be unusually lucky to find the calf by chance.

We say the cows laid out Boston. Well, there are worse surveyors.

– Ralph Waldo Emerson, from
The Conduct of Life

HERE IS JUST A FIELD OF GRASS...

... with red and pink and white clover, blue forget-me-nots and white chickweed, gold celandines, light and dark seed heads, and deeply buried secrets that will emerge in stages throughout the year: pignut, betony, wild thyme, cranesbill, tormentil, lady's smock, orchid. It will be visited by butterflies, like this brown argus in front of me, the colour of magic, its velvet sheen set with jewelled orange–gold marquetry. Not a butterfly to seek, more one to come across when looking for something else. At one moment tiny, still, and happily posing, and then invisible, as its wings close and it becomes part of the bleached grass seed heads it balances on, its underside silver and blue and sparkle.

THE VERY MINUTEST BEAUTIES

While reading Coleridge's essays on Shakespeare – more accurately, the written account of his lectures, compiled from notes taken by his nephew – I come across this:

> Shakespeare possessed . . . deep feeling and exquisite
> sense of beauty . . . affectionate love of nature
> and natural objects, without which no man could
> have observed so steadily, or painted so truly and
> passionately, the very minutest beauties of the
> external world.

He cites the depiction of the hare in *Venus and Adonis*, trying in so many ways to not only outrun but outwit its pursuers:

> *. . . how he outruns the winds, and . . .*
> *He cranks and crosses with a thousand doubles . . .*
> *. . . like a labyrinth to amaze his foes.*
> *Sometime he runs among a flock of sheep*
> *To make the cunning hounds mistake their smell . . .*
> *. . . poor Wat, far off upon a hill,*
> *Stands on his hinder legs with listening ear,*
> *To hearken if his foes pursue him still . . .*

Robert Burns wrote: 'Unhuman man . . . to shoot a hare.'
I reflect with sadness that I haven't seen a hare on this farm for many years.

HUES THAT LIFT THE SPIRIT

In spring the poplars are pink and woolly soft. But come the autumn and they are yellow and bouncing the beachball sun from one to the next, streaks of vital light laughing the season in. Soon they will become pink again, briefly counterfeiting spring, thronging the valley with deceiving hues that lift the spirit.

WINNOWING THE WHEAT

The sheep in the orchard – And Dorable, Daphne and Daphne, Chocolate and Emily Rose – have no self-filling water trough, so every day their two buckets, dark green and pale green, are filled with fresh water. And every single day without exception they drink from the dark green bucket first.

I won't say that all sheep are compassionate, but some are. There have been many instances of sheep exhibiting concern for people as well as for others of their own species. When Ellen, a hand-reared lamb, was just one week old, sitting patiently waiting for her feed time, she noticed me burst into tears for some reason I now forget. She stood up, came over to me and looked up into my face with what I can only describe as concern. Many years later when I was taking hay for her and a few companions, a strong gust of wind blew the Land Rover door forcefully onto my knee and I hopped around in agony. She left her hay and came to stare up into my face and would only resume eating when I convinced her I was fine.

If an individual sheep finds itself at a disadvantage, lame and not able to keep up with the flock, for example, or entangled in a bramble bush, it tends to have a 'friend' hanging around to keep it company. Virgil refers to 'sympathetic sheep' in *The Eclogues*.

Today the sheep are watching on in curiosity as I winnow wheat – and 'today' could be any day of the week, every week, for the past fifty years. My maternal great-grandfather was a miller and farmer. He was born in 1850 and father to eight, all of whom I knew. I'm glad I didn't know him, though. By their reports, he expected his wife and children to work hard while he strutted about dressed like a gentleman doing nothing. The mill house had a door that opened directly onto the terrifying millwheel, which held an addictive fascination for my mother as a small child. I particularly liked the story I was told of the man who delivered sacks of flour every day by horse and cart. He was given his wages before the last delivery on a Saturday and he would arrive back at the mill, drunk and asleep, courtesy of the faithful horse who trudged homewards unguided and reversed the cart up to the steps, while waiting for his handler to wake. I scoop the wheat up and let it fall slowly into a riddle and the wind blows the dust away. There are some 'idle weeds that grow in our sustaining corn'* but I assume they add to its nutritional profile.

And wild yeast, invited in, to perform miracles of leavening and digestibility, slow and sure and eternal. To paraphrase Merlin Sheldrake in *Entangled Life*, perhaps it is not we who have domesticated yeasts, but vice versa.

Miracle of miracles, considering that I have had to eliminate sugar from my diet in order to avoid the

* *King Lear*, Act IV scene iv.

headaches that used to plague me, I have adapted a recipe for a completely delicious, wild-yeast-risen cake, with absolutely no butter and no sugar, and sweetness supplied by many small pieces of apple.

A few months after this I read William Barnes's 'Eclogue: Father Come Hwome' and realise I am not as original as I'd hoped:

> I got a little ceäke too, here, a-beäken o'n
> Upon the vier . . .
> He's nice an' moist; vor when I wer a-meäken o'n
> I stuck some bits ov apple in the dough.

Cooking and eating come into every equation. We have to feed ourselves well and stay healthy in order to be able to look after our animals, and we have to feed them well so they are healthy, and then the produce we supply will make our customers healthy. Everything is connected, and if one tiny link in the chain is dented it will have an effect somewhere else. You can't cheat nature.

John: . . . an' what is woose,
I fear that I must zell my little cow.

Thomas: . . . What, can't ye put a lwoaf on shelf?

John: Ees, now;
But I do fear I shan't 'ithout my cow.
No; they do mëan to teäke the moor in . . .

Thomas: . . . yet I s'pose there'll be a 'lotment vor ye . . .

John: No; not vor me, I fear . . .
Vor 'tis the common that do do me good . . .

– William Barnes, from 'The Common
a-Took In', *Eclogues*

A SPECIAL FRIENDSHIP

Diary entry, 1 December 2019

Another day, and the raven floats overhead to his wood-land bed; it's late and nearly dark. Two minutes later and two buzzards drift down to their roosts fifty yards further down the same wood. Nest-building and drifting they do not 'lay waste their powers' (to steal a phrase from Words-worth).

Gareth and I take big bales of straw for the cattle to lie on and, as ever, it makes me think of winters past, any and every one, and bedding down the cattle with clean, bright straw. Who could not feel tired if asked to stand next to a luxurious, clean and comfy bed? They sink down in ecstasy, forgetting how hungry they were a moment ago.

Some leap up again, unable to suppress their joy and, dangerously, set about helping us to spread the straw. They join forces to roll giant bales, burrow into them and almost disappear, fight, play, kick, jump. Clean straw 'sheets' are better than supper; we sometimes have to hide behind the tractor to avoid injury. For the moment they are contained, yarded, cooped up. But, as soon as they are given their freedom again, they will behave with dignity and restraint.

Just a week before Old Dolly calved we saw a large swelling on her upper rear leg. She didn't appear to be in pain and was eating normally. She calved without trouble but despite the vet's best endeavours the swelling remained and her limp became more pronounced.

Cows can be wonderfully gentle but it has been known for several to gang together to attack a disabled peer, and so Dolly and her daughter were given a field of their own.

When the weather closed in with such constant rain that the whole herd had to be housed, Old Dolly was given semi-detached accommodation next door to Gold Celandine.

Winter progressed and after many weeks and without warning, her calf, by now plenty old enough to be weaned but still suckling milk every day, suddenly left home. She jumped two gates: the first, away from her mother and into Celandine's pen, then out into the big wide world, eventually moving in with the neighbours a few dozen yards away. Old Dolly, to our surprise, was relieved. She had the place to herself and her bossy, conceited, beautiful, difficult daughter had freedom and a social life and no second thoughts.

When drier weather came and all the cattle were back out in the fields, Dolly was given carte blanche: anywhere, anytime but no socialising. She rose to the challenge and used the bits of the farm, garden and barns that were hers to advantage. She took her time, got up late, stayed out late; sometimes we didn't see her at all. She still limped but looked rejuvenated, glossy, proud. Without human interference she showed us exactly how well she could behave.

We find this always happens. If we allow one cow to go for a walk when the whole herd is housed for the winter, she will invariably behave beautifully, sometimes hunting for a few blades of grass, sometimes choosing a particularly nice bale in the barn, sometimes visiting a daughter or friend in a different yard. If we let two go together, and especially if they are youngsters, they will plan and execute some form of mischief and refuse to return when asked. A lone cow will go 'home' politely after a sojourn, having attracted no great attention. If two calves jointly receive a freedom pass, absolutely every other member of the herd is jealous.

While on the subject of Old Dolly, I am reminded of her first daughter's tendency not to mean a word she said. Like a small child repeating one of its parents' mantras without knowing what it meant, daughter Dolly copied her mother's angry shake of the head. In Old Dolly's case, this was a warning to all humans to keep their distance. The younger D was invariably friendly, yet she still shook her head angrily as a greeting, before immediately submitting to being stroked and loving it.

I could say a thing or two about Gold Celandine, the first daughter of the original Celandine. She was born in a thunderstorm and briefly looked ethereal and vulnerable, a tiny, pale-gold calf standing next to her huge black mother with her imposing white 'pitchfork' horns as the rain lashed down and the sky split open with streaks of lightning. But by the time dusk was enveloping the field, she had already had a good feed of milk and was equal to whatever the weather threw at her. By the end of her third day, having been acknowledged by all the other cows, she was a paid-up member of the herd. In time, she too would be a large and imposing cow, gold with large, paler gold horns, and wonderfully astute.

When a calf is born in biting wind or pouring rain, it adjusts its view of the world accordingly. Calves born inside a barn, protected from the elements, are slightly less robust and more susceptible to infections. I assume that whatever the weather they first encounter is the norm to them, though maybe that also happens at a subconscious level in relation to their immune system. Thus, bringing an 'outlier' inside is a less risky decision than turning an inside dweller out. If a calf is surrounded by grass it will know exactly when to try to eat it. If the decision to introduce a calf to grass rests with humans, it is easy to choose the wrong day.

When Gold Celandine cut a teat (on beastly barbed wire I suspect) I could see the gash from afar, with binoculars. Twice-daily calendula ointment pleased her so immensely she began to anticipate my arrival and found ways of showing her gratitude by exhibiting motionless bliss, even though I knew she wanted to kick out from the pain. Lotions made with calendula really do seem to have amazing healing properties and as a family we've come to rely on them ourselves.

Many years later, and only a couple of days before she was due to calve, Gold Celandine looked unwell. I walked her home. She had a high temperature and mastitis. I followed the vet's instructions. She calved, but she had no energy to tend the calf and it slid under the gate, probably before it could even stand up. It then managed to slide under a different gate and ended up in the wrong pen altogether. I was momentarily mystified to see a subdued Celandine in one pen and a new, wet, pale-gold calf in another. Initially I had to milk her three good quarters to feed the calf as she felt too unwell to stand still for it to suckle. She was improving too slowly.

It was wintry and I was busy tending the rest of the herd and flock, but it suddenly hit me that what she needed was not just injections and procedures, but love. I placed my hands on her big body and could feel her whole being relax. Neither of us moved. After a while, I gently stroked her and then, reaching for the brush, I groomed every inch. This became a daily ritual and she steadily got better. On a lazy summer day, it would have been my first, instinctive reaction, but I had allowed my tiring winter schedule to dull my thought processes. She never forgot, and we had a special friendship ever after.

LIFE LESSONS

My mother was a wonderful infant teacher and truly loved every single child she taught. She revelled in finding out what the children were good at, even if it wasn't in the curriculum. One boy, who found it impossibly hard to learn anything at all, spent his days making the most magnificent and accurate drawings of motorbikes. Mum was able to encourage and nurture his talent, which helped him feel the confidence to learn everything else too. She didn't teach me and Rich much, though. She seemed to think we would absorb, by some sort of undefined magic, everything she knew.

One thing we *were* taught by both our parents was how to shovel muck. Between 1953 and 1974 the cows were milked twice a day, and the milking parlour had to be kept spotlessly clean. We have aged in tandem and these days our shovelling would probably appear slow to an onlooker, but we feel the same as always and neither of us would admit to tiredness before the other so we can still achieve a great deal.

Almost every day on an extensive livestock farm is hard though generally satisfying work, but there are many things that I can't write about: the really stressful or the truly heart-breaking or the intimate. TB tests, as an example of the first category, would be impossible to describe. The test alone means nothing if it's not set in its turbulent historical, medical, political and emotional context.

WHAT'S IN MY LAND ROVER

Bucket, or preferably two
Dustpan and brush for cleaning water troughs
Five-gallon camping container for water
Bolt cutters
Wire cutters
Claw hammer
Staples and nails
Thermometer
Walking stick
Spanners, adjustable and others
Screwdriver (though no screws!)
Saw
Drill, from time to time
Chainsaw, more often
Thistle stocker
Strong trowel, for rescuing tiny trees before the
*   mower or topper enters any field*
String, lots of string
Rope, if I'm lucky enough to be able to find any;
*   however often I buy it, someone else always*
*   'needs' it immediately*
Binoculars
Driving glasses, reading glasses, notebook/diary and pen

Bottle of water for me
Plumbing fittings for troughs with ball valves and
  floats
Halter

If it's spring and lambing time then the list is much longer. There is a penknife in my pocket at all times, no matter what. It has saved lives.

LEARNING LESSONS

Sheep have a deep suspicion of wide-open gates and understandably so. To achieve our goal of bringing them into the yard, we have to remember to leave the gate just a tiny bit open, as if by mistake, and only then will they all nip through, feeling naughty and elated.

One eight-year-old sheep, old Cocoa, who has daughters aged six and four plus attendant grandchildren, always waits for them before she goes out to graze. If they go out before she does, she talks to them in a very low, deep voice, asking them to wait for her. Lambs learn lessons faster than calves, I think, and when the hand-reared lamb Theophilius had to share his lawn with old Cocoa one spring, he watched intently as she grazed what he had only played on, and then followed suit.

The game of chess that is farming demands we think several moves ahead before moving at all. No matter how long or glorious the summer, we always plan for winter, and in winter we plan for spring. But we also have to factor in the concept of 'clean grazing' to avoid specific stomach problems developing in young lambs, and remember that the bull needs to be at least two fields away from heifers that are too young to go in calf yet and any other females to whom he is related. Add in the fact that all the fields and all the groups of animals are of different sizes, and things can get very complicated.

HOW TO NAVIGATE A FIELD OF COWS

Diary entry, 17 July 2020

We are mowing several fields of grass in the hope of making hay if the weather holds. I sit in the wildflower meadow with grasshoppers and crickets exploding from leaf to leaf all around me. Busy spiders weave webs to catch their suppers. Tiny beetle-like creatures crawl and hurry at ground level, hard to see. The wind makes incomparable music among the wildest flowers and most ancient grasses and in the towering trees all round me. I wonder if wind is silent where there are no trees. The grass is probably noisy at other times of year but I sit and stare at it only when it is warm and dry and beckoning.

From the elevated vantage point of his tractor seat, Gareth can see a small group of walkers seemingly stopped in their tracks by a large bunch of cattle. He mows in their direction, stops the tractor and gently encourages the cattle to move away. The walkers continue down the footpath. I think of the entry in Dorothy Wordsworth's journal for Thursday, 18 March 1802: 'I went through the fields and sate half an hour, afraid to pass a Cow. The Cow looked at me, and I looked at the Cow, and whenever I stirred the Cow gave over eating.'

The subject of how to walk through a field of cows

safely is not one I would have considered until a few years ago when I read a headline telling me someone had been killed by cattle. Before that day, I would have thought it impossible, but since then I have indeed been asked several times how to navigate a field of cows safely.

I can never supply a fully satisfactory answer, partly because every incident will be different, but mostly because I do not walk through fields of other people's cows. There are points to make though: it is entirely understandable that the public should wish and expect to be able to walk unassailed by dangerous animals, but all animals require peace and safety from being disturbed by humans too. There are ideas and snippets of advice I could give, but none would be foolproof or cover every situation. I am inclining towards the belief that a tweaking of the footpath network to enable walkers to avoid danger and animals to avoid disturbance might be achievable. Some modern technology, doubtless in the form of an app, could perhaps enable farmers to record fields where possible danger lurks and to suggest temporary alternative routes. This might also be used to save sheep being chased and attacked.

When I was growing up, almost everyone I encountered had some link with farming, through a parent, grandparent or some other relation or friend, and because of that they had some intuitive or learned understanding of how to behave and how to avoid potentially dangerous situations. Today, the situation has been reversed. Yet the desire to have some land of one's own and keep animals and grow food is rightly and thankfully still there in a significant proportion of the population. The problem is that agricultural policies and economics, with food prices far lower in real terms today than they were in the 1950s, force farmers to increase the size of their farms to remain economically viable, which means that less and less land becomes available for those who would like to produce their own food on a smaller scale. Having no control over the production of the food you eat is a significant and wide-reaching problem. When the enclosures deprived people of even the tiny strips of land they previously had, and the industrial revolution and the introduction of machinery were making agricultural jobs redundant, those who could not feed themselves ended up in towns. Food had to be bought, and thus a very small number of people were handed unprecedented power, and with it the temptation to adulterate and dilute staple foods to make more profit. A tiny number of food manufacturing giants came to have control over the diet and therefore the health of the majority of the population. And since everyone has to eat, control over the food supply is simply a licence to print money.

I read that access to green spaces has been proved to benefit people's mental health and improve children's academic achievements. Having the power to grow things to eat could alleviate, if not solve, multiple crises: obesity, depression, diet-related illnesses, and all the ills that are caused by lack of exercise and loneliness.

236

... sich my happy feäte is,
That I can keep a little cow, or ass,
An' a vew pigs to eat the little teäties.

– William Barnes, from
'The 'Lotments', *Eclogues*

TASTING THE PAST

The first half of 2020 saw me largely out of action due to a bad leg wound (a long story involving a sheep rather than a cow this time) but I was chauffeured round the farm whenever anyone had time to take me. First stop: the new 'old' orchard, a one-hectare stand of recently planted old varieties with local significance. I watched as Gareth picked small pretty apples smelling and tasting of strawberries: Gladstone. And very small, dark-blue plums: Rivers Early. It is exciting to be able to taste the past this way.

My maternal great-grandfather had three orchards with every conceivable variety of apple, pear and plum, each one picked with enormous care and reverence at precisely the optimum time. As an old, traditional apple- and pear-grower in the Vale of Evesham drummed into me, 'All top fruit has two ripenings. There's the time they are fit to pick and the time they are ready to eat, and they are not the same.' He added that if more people knew that and knew when each variety comes ready to eat, a lot less garden and small-orchard fruit would be wasted and a lot less would need to be imported.

My mother never forgot the terror she felt at being sent up the never-ending pear-tree ladder, as she was small and light and unlikely to break a bough. Oddly enough, I came across a photograph in a fascinating catalogue entitled 'Warwickshire Women: A Guide to Sources' in the county record office, of two women washing high windows, standing on the top rungs of pear-tree ladders. During the Second World War, the War Agricultural Executive Committee (Warag) ordered the removal of every one of the fruit trees on the farm my maternal grandfather managed at the time, to make way for wheat. This necessitated gangs of men with saws, dynamite to loosen the roots, and giro-tillers to remove them, but the orchards were small and unsuitable and they produced tiny quantities of wheat. At the same time the local population was deprived of fresh and nutritious fruit, and chafer beetles – among others – lost their vital habitat.

We leave the orchard, driving to the highest part of the farm. The sun reappears, looking as if he's run out of fire, and reminding me of John Clare describing him as 'beamless and pale and round, as if the moon . . . Had found him sleeping and supplied his place'. But by the time he is ready for bed, he has regained a gorgeous glow and lights up the trees all round the valley before sinking to sleep behind the furthest hill.

TIME IS THE ONLY THING ANY OF US OWNS

Later that summer, my leg was somewhat better and I could walk short distances. One evening I entered the wood's cool, and found the perfect spot to sit and observe.

There are many ways of getting 'lost' in this wood but all of them require a little effort. The trees grow on a steeply sloping, north-facing former field, where one of our predecessors grazed his Ayrshire cows, and most points of entry require a climb. For my own safety, I reluctantly decided to risk crushing part of the carpet of wood sorrel – flat-footed of course, to do least harm – and discovered to my delight that it extended far further than I knew. The trees are mostly Douglas fir and redwood cedar, and since they were planted in the early 1960s they have grown immensely tall. The sun finds ways to get in between the tightly packed trees, reminding me of the shaft of light that shines through Brunel's Box Tunnel near Bath. The fairly bare woodland floor was dappled with patches of ferns, mosses and tiny ash saplings, and even an occasional surprised frog. Roe deer patrolled calmly, wishing humans did not exist so they could once again own the pastures; as Peter Wohlleben writes in *The Secret Network of Nature*, 'Deer have a love–hate relationship with trees. They don't actually like forests.'

Despite the less than ideal conditions created by the pine interlopers in this Cotswold corner, a large number of ash trees grow to maturity here, and I secretly hope the wood will one day revert to a deciduous state. The sun was hot on the open fields but here it was cool and quiet. I cannot enjoy the wood equipped only with a beak and a brain, as a bird can, so I carry a small bag with binoculars, a notebook and two pairs of spectacles, one for looking and one for writing.

A mewing buzzard was talking in the air above me and I glimpsed it intermittently in the blue gaps in the woodland's roof. The kites are in residence here too, after a long absence of perhaps a century and a half. Humans take away and humans decide when or if to give back – often because they can rather than because they should. Kites were hunted to extinction and then twenty or so years ago a few pairs were brought from Spain and re-established in the Chilterns, where they bred and multiplied and expanded their territories until they came 'home'. If we go to the top of our highest hill we can watch their floating acrobatics from above. From the valley we see them only from below.

I had come that day with a goal: to see what I see. The promise of a goldcrest sighting made me happy to carry the heavy binoculars. Do any other species watch one another for pleasure, I wonder. Humans are not sufficiently self-contained. Every other creature has an occupation and pursues it diligently. Only people have manufactured the oddest commodity: spare time. Time is the only thing any of us owns, and the only thing of any real value that we can give to others. If we choose not to give it, we end up inventing ways to use it, giving ourselves endless choices and decisions and often creating stresses where none existed before.

Hunters have to be patient and speedy, and the hunted have to be watchful; neither wastes time wondering what to do next.

I sat in a circle of sunlight and waited. Butterflies flew past: tantalising orange flashes, white–green streaks, ribbons of indistinct colour.

Sitting quietly in the wood I recalled the archive section in *The Times* that I had read recently, about Mr John Grimshaw Wilkinson, the blind botanist who discovered in the early twentieth century that during heavy showers trees were distinguishable by the sounds they made. The oak was the noisiest and the Scots pine the quietest. The opening words of *Under the Greenwood Tree* show that Hardy knew this too: 'To dwellers in a wood, almost every species of tree has its voice as well as its feature.' Roger Deakin refers to this in *Wildwood* and in Virgil's *Eclogues* we find, 'It is the shepherd and his sheep that are her [Nature's] confidants. It is they who comprehend when the woods . . . make music and the pine trees speak.'

Eventually I emerged from the wood and sat on the crushed-stone roadway in the paddock. The littlest Daphne joined me, settling herself down catlike by my side. Her older sister was three yards to our right, leaning on the gate I had come through.

We sat companionably. The water-mint-filled stream created its own small breeze as it rushed past. A buzzard swooped low and glided just over our heads. The funnel-shaped trap of a spider's web caught my eye, tucked in between two small stones and half hidden by a small wild geranium with leaves resembling nothing so much as a hawthorn. I anticipated William Keble Martin's *Concise British Flora* telling me there is a hawthorn-leaf geranium – it didn't. Daphne wandered off to nibble anything green, followed a minute later by her sister; it must have been nibbling time. I sat a few moments longer and counted how many species of tree I could see without moving: lime, elder, rowan, white and Lombardy poplar, willow, sycamore, maple, cherry, walnut, ash, silver birch, beech in the far distance. Although I could not see them, I knew that the tall, ungainly red oaks were fighting for their share of light in the woodland, scarcely twenty yards away.

The wind rolled the surface of the lake and a turquoise damselfly darted among the water mint, three feet high and with flowers to rival any orchid. The lake's 'beach' of mint, meadowsweet and hemp agrimony, bordering a wide band of mare's tail punctuated with willows, supplies infinite secret caverns for dragonflies and butterflies and many less beautiful and often more dangerous insects: horseflies, wasps, bees that inhabit old oak fence posts and don't make honey, and ordinary-looking flies who must have a purpose but I haven't yet discovered what. Slowly from stage left, a cow grazed into view, eventually coming to a standstill reflected in the lake, upside-down.

OLD HENS AND NEW TRICKS

I found myself the other day standing under a tall, young ash tree festooned with baby long-tailed tits, happily oblivious to my presence. Red kites were rising and falling, rising and falling, tinted orange and gold by the sun. They alighted softly onto the short sward and sat like contented seagulls floating on a green sea. I went for a last look round the heavily pregnant cows just before dark and was suddenly aware of a young but fully fledged buzzard on the fence post beside me. It hesitated, then skimmed twenty feet of field to wobble on a hawthorn tip too frail for its weight. I reached the gate and it quickly returned to the post. I had a strong feeling that its leaving-home instructions mustn't have included the address of the nearest safe resting place.

Woods make their own wind and the sheep evidently appreciate the moving cool of the small beech wood. I found a seven-week-old lamb on her own by the top wall. I approached, and she got up and limped slowly and painfully away from me. I returned to the farm for Gareth and we encircled her with soft netting before conveying her home. We kept her in the garden, and she quickly learned to suck milk from a bottle. We followed the vet's advice for ten days, during which time she spent most of each day in the corner of the garden nearest to the field we took her from, staring at the garden door. I was amazed at her sense of orientation.

I tried to win her confidence as, unlike a genuine orphan lamb, at seven weeks she was almost grown up and totally 'immune' to humans. I still had to catch her in order to feed her, although immediately après milk she was, briefly, devoted to me. I carried her into the kitchen one day, where it was cooler. I sat on the floor and sang to her, thinking of Orpheus and his ability to charm all living things and even stones with his music. Gareth nodded off first and then the lamb. After that, we were firm friends.

Once her course of treatment was over, I returned her to the field and placed her next to her mother and siblings; she was the smallest of triplets. Her brother walked up to acknowledge her and they grazed away together into the evening light. Happy, I left them to it.

Next day, however, and on subsequent thrice-daily visits to take her milk, I found her with different sheep and began to doubt if she was remembered. Each time, she would listen out for my vehicle and race over, calling loudly. As ever, fortunately, her adoration lasted roughly one minute before the grass beckoned.

We had recently been given a new hen. She was two years old; our old pair were both nine. Initially we gave her a pen within their enclosure in case they fought, but after they had been talking to one another through the wire mesh for a week, we opened the gate and watched. The old ladies were more than twice as large as the tiny newcomer but she pre-emptively pecked each of them quite hard on the top of their heads and then rushed around, excitedly exploring and seed-pecking. In one second, she had established herself as the boss and they fell into line. The beautiful corollary was that her lively presence completely transformed them. From living very definitely in retirement mode, pecking half-heartedly at their food and consuming tiny quantities, they became quick and active, began to eat keenly and took a renewed interest in everything.

WHAT TIME DO BATS GET UP?

The other evening I vaulted the gate (sounds better than clambering over) into the L-shaped Field and was immediately chaperoned by a lone bat, who accompanied me in friendly circles as I walked diagonally from gate to gate, before being absorbed back into the night. Fifty paces later I met a lone moth . . .

TENDERNESS AND TRAGEDY

You don't have to be friends with a bull, but you do need to live your life on his terms.

With Jake it was easy; from just a few weeks old he would walk up to me and tug my clothes with his teeth, asking for attention as surely as a cat seeking affection by trying to trip you up. Jake stayed true to his sweet temperament and trusted us completely. Even if he couldn't immediately see the benefit in doing something, he knew for certain that things would end favourably. We resolved we would never disappoint him or ask him to do anything unnecessary, so that should we ever need to insist he do something, he would be inclined to agree.

I decided to speak to our more recent arrival, Prometheus, every day, and indeed every time I saw him, but never in a manner to invade his privacy. I needed him to recognise my voice and yet ignore me. I knew there might be occasions when I would have to ask him to do something or go somewhere, on the understanding that if he obeyed, he would always end up somewhere better. In the meantime, I needed to be able to enter his space as a mere presence, so I could work, or observe, as unobtrusively as possible.

In September 2020 he sprained his ankle, or so I thought, and very sensibly removed himself from the herd. In fact, it took considerable effort to find him, altogether hidden from view on a small patch of land with very tall ferns behind him and immense blackberry bushes in front and to the sides. I moved the rest of the herd into an adjoining field and set about waiting on him.

He understood exactly what I was doing and approved. His usual aloof expression softened and he gently ate hay and willow from my hands. His tug was strong and he pulled me so close I almost fell into his 'lap'. After this I piled the hay high and tied the willow firmly so he could tug the fence. Carrying water for his huge thirst was quite an ordeal and Gareth came to my rescue, heaving heavy containers of water over the fence, filling buckets big-enough-for-a-bull's-big-head to place in front of him, and then repeat. Prometheus stayed in that spot for two days and then I had another search mission. He had sought out the only willow tree in the area and had munched every reachable leaf. He watched me approach and climb into the tree to break more boughs. This time his gratitude was palpable. He munched and munched and then stopped; cattle always know when to stop. Rest, water, willow, hay and then grass as soon as feasible. It will take a few weeks to repair.

Not so; this was to signal the end of Prometheus. It was an odd coincidence of timing that we had just bought Prince, to mate with the older bull's daughters. I wanted his injury to be curable and hoped that by allowing him infinite rest it would be. Richard persuaded me to summon the vet who could tell it was his hip that was injured. We had to make a decision based on his welfare, and with heavy hearts we made an appointment for him to be killed. It was enormously sad for us, but we managed to make it calm and maybe, odd though it might sound, even enjoyable for Prometheus.

I had envisaged his last day on the first day he arrived here, driven from Lincolnshire by the man who bred him, in a transporter he was accustomed to riding in from field to field from a very young age. He had travelled well: content, calm, clean. He stepped out with a look of disbelief that clearly said, 'Is all this mine?' before trotting off to investigate. I hoped then, at that moment, that we would keep him forever but knew that would not be possible. I found myself thinking that I hoped he would not forget, however many years elapsed, what a transporter was. He didn't. I tied fresh boughs of ash in the front, then crawled in through the small side door, slicing sweet apples in half so he could smell them. The enticement worked, and this enormous giant stepped carefully on to the ramp and started munching. We felt dreadful. Richard drove carefully and, marvellously, even on arrival at the abattoir Prometheus tiptoed out enthusiastically, looking round once more at a new, unfamiliar and very temporary kingdom. All previous journeys had ended somewhere better and his end was as good as it could be.

As farmers, we know the risks and make our choices, and when we suffer we understand why. It's not okay for the animals in our care to suffer. There are moments of tenderness to be savoured and of tragedy to be learned from, or simply endured.

THE KEEPING OF SECRETS

I am perching on the cliff edge of the long-abandoned quarry – filled in now with hundreds of years' worth of beech leaves and adorned by the exposed, limb-like roots of the trees – admiring the sheep, who rarely make use of this 'beechen green' shade. The heat of the sun is filtered by the still-green leaf canopy, and the sheep, dotted around on precarious ledges, are fanned by cooling breezes and un-troubled by the flies that are plaguing the half of the flock that chose instead to cluster in the scant shade afforded by the stone wall and boundary trees. This beech wood, which both softens and stabilises the old quarry, creates its own climate: frequently too cold for cattle or sheep but occa-sionally balmy and air-conditioned.

A few months ago I found a plant here, sporting a rosette of seven leaves on a single stem. It reminded me slightly of Herb-Paris, but that rarely has more than four leaves. I was hoping a flower would appear eventually to aid identification but luckily a friend recognised it from a photo I sent her: the leaves of the green hellebore.

Every year since we've farmed at Kite's Nest I have found new plants, tucked in behind fence posts, hiding behind trees, and so obviously out in the open I must have walked past them a hundred times without noticing. I long to see and find everything, but part of me knows how extremely important it is that the farm keeps as many secrets as it can.

ACKNOWLEDGEMENTS

I would love to be able to say that I was 'discovered' by Faber & Faber but I cannot. I believe I can say that I was 'found', though. Laura Hassan republished *The Secret Life of Cows*, sending that little book out across the world, and me on a wild and wonderful journey with it. Without her kindness, professionalism and support, this companion volume would not exist.

I have also been unbelievably lucky that my words have been scrutinised and shaped by Hannah Knowles, Fred Baty, Josephine Salverda, Jill Burrows and Justyna Bielecka.

The text is gorgeously illustrated by Joanna Lisowiec, for which I am delighted and grateful.

Thanks also to the Sustainable Food Trust for permission to use parts of articles originally published in their online newsletter: www.sustainablefoodtrust.org.

Numerous friends who believe in everything we do at Kite's Nest, and whom I dare not list for fear of omission, mean the world to me; they know who they are.

My brother Richard and my partner Gareth are my constant guiding lights. We are all so different yet we share the one most vital conviction that the animals in our care come first. Richard and Gareth take over seamlessly when I am writing or otherwise out of action, and it is both glorious and a little worrying that they can cope so easily without me.

BIBLIOGRAPHY AND
FURTHER READING

Barkham, Patrick, *The Butterfly Isles: A Summer in Search of Our Emperors and Admirals*, Granta Books, 2011

Barnes, William, *Selected Poems*, Penguin Classics, 1994

Bersweden, Leif, *The Orchid Hunter: From Irish Lady's Tresses to Shakespeare's Long Purples: A Summer in Search of All Fifty-two of Our Native Orchids*, Short Books, 2017

———— *Where the Wild Flowers Grow: My Botanical Journey Through Britain and Ireland*, Hodder & Stoughton, 2022

Callow, Philip, *Louis: A Life of Robert Louis Stevenson*, Constable, 2001

Clare, John, *The Shepherd's Calendar*, Oxford University Press, 1973

Coleridge, Samuel Taylor, *Coleridge's Essays and Lectures on Shakespeare and Some Other Old Poets and Dramatists*, J. M. Dent, 1907

Conford, Philip, *The Origins of The Organic Movement*, Floris Books, 2001

———— *The Development of the Organic Network: Linking People and Themes, 1945–95*, Floris Books, 2011

Fiennes, William, *The Snow Geese*, Picador, 2002

Flannery, Tim, *Europe: A Natural History*, Allen Lane, 2018

Grahame, Kenneth, *First Whisper of 'The Wind in the Willows'*, Methuen, 1944

Grindrod, John, *Outskirts: Living Life on the Edge of the Green Belt*, Sceptre, 2017

Hayes, Nick, *The Book of Trespass: Crossing the Lines that Divide Us*, Bloomsbury, 2020

James, Jeremy, *The Byerley Turk: The True Story of the First Thoroughbred*, Merlin Unwin, 2005

Keble Martin, William, *The Concise British Flora in Colour*, Ebury Press, 1965

Lampkin, Nicolas, *Organic Farming*, Farming Press, 1990

Lindén, Axel, *On Sheep: Diary of a Swedish Shepherd*, Quercus Editions, 2019

Littlewood, Joan, *Joan's Book: Joan Littlewood's Peculiar History As She Tells It*, Methuen, 1994

Percival, Rob, *The Meat Paradox: Eating, Empathy and the Future of Meat*, Abacus, 2023

Poux, Xavier and Pierre-Marie Aubert, *An Agroecological Europe in 2050: Multifunctional Agriculture for Healthy Eating (Findings from the Ten Years For Agroecology (TYFA) Modelling Exercise)*, Institut du Développement Durable et des Relations Internationales, 2018

Rebanks, Helen, *The Farmer's Wife: My Life in Days*, Faber & Faber, 2023

Rebanks, James, *The Shepherd's Life: A Tale of the Lake District*, Allen Lane, 2015

——— *The Illustrated Herdwick Shepherd*, Particular Books, 2015

——— *English Pastoral: An Inheritance*, Allen Lane, 2020

Squires, G. A., 'The Small Farmer on the Land', in Harold John Massingham (ed.), *The Small Farmer: A Survey by Various Hands*, Collins, 1947

Street, Arthur George, *Farmer's Glory*, Faber & Faber, 1948

Sverdrup-Thygeson, Anne, *Extraordinary Insects: Weird. Wonderful. Indispensable. The Ones Who Run Our World*, Mudlark, 2020

Tinsley, Anthony Brian, *Horse and Cart Days: Memories of a Farm Boy*, Clark & Howard, 1990

Tree, Isabella, *Wilding: The Return of Nature to a British Farm*, Picador, 2018

Trevelyan, Marie, *Glimpses of Welsh Life and Character*, John Hogg, 1893

Treverton-Jones, Tamsin, *Windblown: Landscape, Legacy and Loss – The Great Storm of 1987*, Hodder & Stoughton, 2017

Wayland Barber, Elizabeth, *Women's Work: The First 20,000 Years – Women, Cloth and Society in Early Times*, W. W. Norton, 1996

Wohlleben, Peter, *The Secret Network of Nature: The Delicate Balance of All Living Things*, Bodley Head, 2018

Wordsworth, Dorothy, *The Grasmere and Alfoxden Journals*, Oxford University Press, 2008